MICROARRAY METHODS AND PROTOCOLS

MICROARRAY METHODS AND PROTOCOLS

Robert S. Matson

CRC Press
Taylor & Francis Group
Boca Raton London New York

CRC Press is an imprint of the
Taylor & Francis Group, an **informa** business

CRC Press
Taylor & Francis Group
6000 Broken Sound Parkway NW, Suite 300
Boca Raton, FL 33487-2742

First issued in hardback 2017

ISBN 13: 978-1-138-40699-5 (hbk)
ISBN 13: 978-1-4200-4665-6 (pbk)

Library of Congress Cataloging-in-Publication Data

Microarray methods and protocols / edited by Robert S. Matson.
p. ; cm.
Includes bibliographical references and index.
ISBN-13: 978-1-4200-4665-6 (hardcover : alk. paper)
ISBN-10: 1-4200-4665-9 (hardcover : alk. paper)
1. DNA microarrays--Laboratory manuals. 2. Protein microarrays--Laboratory manuals. I. Matson, Robert S. II. Title.
[DNLM: 1. Microarray Analysis--methods--Laboratory Manuals. 2. Analytic Sample Preparation Methods--Laboratory Manuals. 3. Gene Expression Profiling--methods--Laboratory Manuals. 4. Laboratory Techniques and Procedures--Laboratory Manuals. QU 25 M6258 2009]

QP624.5.D726M513 2009
572.8'636--dc22 2008035383

Visit the Taylor & Francis Web site at
http://www.taylorandfrancis.com

and the CRC Press Web site at
http://www.crcpress.com

Dedication

To my mentors: SeaBong Chang, Tokuji Kimura, T.T. Tchen, and Armand Fulco

Contents

Preface

The concepts that define *microarray* were developed in the late 1980s. Less than a decade later we witnessed large-scale efforts to commercialize microarrays, and today they are seasoned tools eagerly employed by a world of scientists. This multiplexing technology (formatted as slides, plates, biochips, or beads) now is undergoing the scrutiny required for the standardization that is essential to drive its adoption in future prognostic and diagnostic applications. Microarray technology continues to evolve, taking on different forms: Originating with the glass microscope slide and biochip, it is now pressing onward into the nanotechnology frontier. From the spotting of cDNA and the *in situ* synthesis of oligonucleotide arrays now come microarrays comprising proteins, carbohydrates, drugs, tissues, and cells. For myself, it has been a great adventure into the multidisciplinary approach to research. I am indebted to Jim Osborne, who championed our cause from the helm of Beckman Coulter's Advanced Technology Center, and to Ed Southern, who inspired us all. I would also like to thank my co-authors for their hard work and dedication in providing such excellent contributions to this book.

I remain hopeful for these arrays of small spots, with the expectation that microarrays will ultimately reduce the cost of healthcare. I invite you to join us in that endeavor.

Robert S. Matson, Ph.D., F.A.C.B.
Orange, California

Editor

Robert Matson, Ph.D., F.A.C.B., has been involved in microarray technology research for the past 17 years at Beckman Coulter. He participated in NIST's Advanced Technology Program–sponsored Genosensor Consortium, and collaborated with Sir Edwin Southern on the development of an *in situ* oligonucleotide array synthesis platform for the corporation. Other work included development of a microplate-based array platform for multiplexed micro-ELISA. Dr. Matson has been granted 12 U.S. patents, as well as several European patents on nucleic acid and protein microarray technology. He was inducted into Beckman Coulter's Inventors Hall of Fame in 2006, and was recently elected a Fellow of the National Academy of Clinical Biochemistry. He has previously served in several technical management roles including R&D director, BioProbe International; R&D director, Costar–Nuclepore; and chemistry R&D group leader at BioRad Laboratories.

Dr. Matson received his Ph.D. in biochemistry from Wayne State University. Following postdoctoral studies at the UCLA Medical School he served as a principal investigator with the Veterans Administration Medical Center and adjunct professor of biological chemistry at the University of California–Davis Medical School. Dr. Matson has also held a faculty lectureship in USC's Department of Chemistry, and was assistant professor of chemistry at the University of Southern Maine, Portland. He is the author of *Applying Genomic and Proteomic Microarray Technology in Drug Discovery* (CRC Press, 2005).

Editor

Authors

Todd Martinsky, cofounder of TeleChem International, Inc., and ArrayIt Corporation, previously served as director of education and consulting at the Codd and Date Consulting Group. Mr. Martinsky has led the ArrayIt Division to play a significant role in the microarray industry. He has authored several book chapters and other scientific literature and has become an internationally recognized lecturer, writer, consultant and teacher. In addition to providing consulting services, Mr. Martinsky has spearheaded ArrayIt's technical support team since 1997. Along with his daily technical and business direction of the ArrayIt product line, Mr. Martinsky established successful alliances with corporate partners in manufacturing, reagents, equipment, and distribution. He is responsible for an educational outreach program that ensures that the broadly patented ArrayIt Micro Spotting Devices are applied in the field with optimal scientific and technological accuracy. He is currently serving on the panel that is crafting future regulatory requirements for microarray manufacturing for the U.S. Pharmacopeia.

David Miklos, M.D., Ph.D., assistant professor of medicine, Stanford University, is a hematopoietic stem cell transplant (HCT) clinician with special interest in chronic graft versus host disease (GVHD). His lab has developed protein microarray technology to measure allogeneic antibody development after allogeneic transplantation. Dr. Miklos earned his B.S. from the University of Notre Dame and his M.D./Ph.D. from Yale University. He trained in internal medicine at Brigham and Women's Hospital, followed by hematology and oncology fellowship training at Dana-Farber Cancer Institute with special emphasis in BMT and hematologic malignancies. He remains active as both a BMT clinician and basic-science immunologist.

Yong Song, M.D., Ph.D., is product manager, strategic marketing, Beckman Coulter, Inc., currently managing the personal cytometry analyzer business. Dr. Song obtained his M.D. from Shantou University Medical College, China, and his Ph.D. from the University of Hong Kong. He received his postdoctoral training under Prof. Melvin Silverman at the Canadian Institutes of Health Research group in membrane biology in the Faculty of Medicine, University of Toronto, Canada. Prior to his current position at Beckman Coulter, Inc., Dr. Song held senior scientific and R&D management positions at Bio-Rad Laboratories, Hyseq Pharmaceuticals, and Beckman Coulter, Inc. Dr. Song has many years of experience in drug discovery and product development for immunoassays and flow cytometry applications. He has more than 80 publications and is an inventor of a multiplex bead-based assay patent for determination of cellular protein modifications, including protein phosphorylation.

Persis Wadia, Ph.D., received her B.Sc., M.Sc., and Ph.D. from the University of Mumbai, India. She is presently a postdoctoral scholar in the Department of Medicine, Division of Blood and Marrow Transplantation, at Stanford University. Her main fields of research have been oncology and immunology. Her current research interests include identifying minor histocompatibility antigens/biomarkers after bone marrow transplantation in AML patients.

Denong Wang, M.D., Ph.D., senior research scientist, Department of Genetics, Stanford University School of Medicine, is specialized in the areas of carbohydrate antigens, anti-carbohydrate antibodies, and carbohydrate microarray technologies. His group published the first description of a carbohydrate-based microarray technology in the March 2002 issue of *Nature Biotechnology*. His recent efforts have focused on identification and characterization of immunogenic sugar moieties of microbes and human cancers, as well as development of novel platforms of bioarrays. Dr. Wang is principal investigator and director of the

Stanford Tumor Glycome Laboratory, which is one of the seven key laboratories of the National Institutes of Health Alliance of Glycobiologists for Detection of Cancer and Cancer Risk.

Masao Yamada, Ph.D., is director of the Glycomics Research Laboratory, Moritex Corporation, Yokohama, Japan. He received his Ph.D. in electrical and electronics engineering from Nagoya University. He has held several technical management positions in his career including director, Advanced Device Development, Fujitsu, Ltd., and vice president, Nippon Laser and Electronics Lab. Dr. Yamada also served as a consulting associate professor for the Solid State Electronics Lab, Stanford University, as well as visiting professor at Nagoya University. In addition to his role as director of the Glycomics Research Laboratory, he is senior general manager for Moritex Corporation.

Chapter 1

Introduction to Microarray Technologies

Robert S. Matson

Contents

Microarrays

It is not the purpose of this book to provide the reader with a detailed account of microarray technologies. I refer you to my companion book, *Applying Genomic and Proteomic Microarray Technology in Drug Discovery* (Matson, 2005) for such detail. You will also find more information within the introductions to each chapter. Nevertheless, we will summarize here the fundamental concepts of and current practices in the field of microarrays. Because this book is primarily concerned with how to make and use microarrays, we will not dwell too much on commercial offerings. However, one must realize that there are certain advantages to using commercial microarray products, especially in the area of quality control and product consistency. Yet, there is good reason to produce your own microarrays if you are interested in analyzing for an analyte that is not available on a current product offering. This is particularly true in working with protein microarrays or other biomolecules such as carbohydrates, lectins, or small organic molecules.

Making Use of Small Spots

All microarrays share a common format based on analysis of data derived from affinity capture of biological molecules by ligands. These are tethered to a solid-phase support (whether it be planar, hydrogel, or bead) confined to a very small spatially addressable region of the support. The simplest embodiment is to immobilize the ligand as a small spot on a glass microscope slide. The key feature of microarrays is that multiple ligands are immobilized that capture different biomolecules (i.e., the array), thereby allowing for the simultaneous analysis of a large number of analytes. Thus, microarrays offer a high order of multiplexed analysis that is not readily available by other means. The most common application for microarrays remains differential gene expression analysis (Figure 1.1), whereas other areas of intense use involve analysis of single-nucleotide polymorphisms (SNPs) and array-based comparative genomic hybridization (aCGH). Since completion of the Human Genome Project, there has been a shift from using cDNA to 70mer oligonucleotides as capture ligands for gene expression work involving humans or cells of human origin. Other genomes are being sequenced now at an incredible rate so that an entire organism's genetic makeup can be queried from a single microarray slide. This brings up

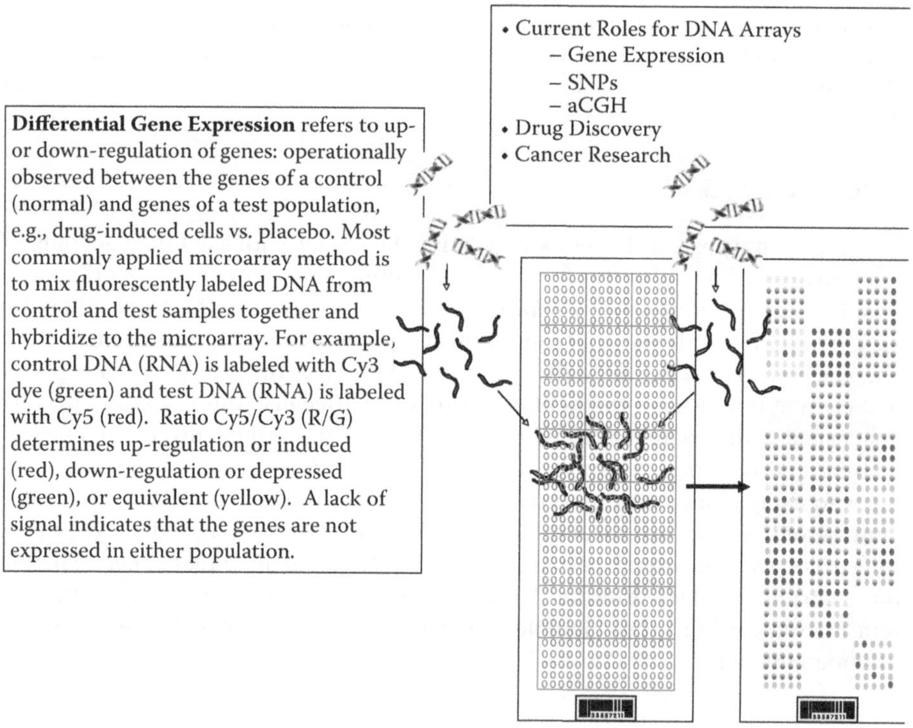

Figure 1.1
Differential gene expression. (Adapted from Freeman, W. M. et al., 2000.)

an important point when considering the use of microarrays: content. You can only analyze for what is printed down on the slide, commonly referred to as "content." It is also important that the content be accurately represented by the oligonucleotide sequence or by the capture antibody in the case of protein microarrays.

The Human Proteome Project (under the guidance of the Human Proteome Organization, HUPO), which was launched following completion of the Human Genome Project (HUGO), is an effort to unravel the human proteome. It is a worthy undertaking since much of our prognostic, diagnostic, and therapeutic medicine is based on understanding protein functionality. Fortunately, our expectations are now tempered with the realization that this endeavor presents a much bigger problem than sequencing DNA. Currently, there are a number of different approaches being taken to address the proteome: the classic frontrunner, 2D-gel electrophoresis; mass spectroscopy; and protein microarrays. There are, of course, pros and cons to each of these technologies, as well as others being developed for proteomics. As far as protein microarrays are concerned, perhaps the biggest drawback is the availability of content. It is estimated that there are perhaps 100,000 transcripts and perhaps up to 1 million different proteins in the proteome if you consider isoforms and post-translational modified proteins. These occur in nature at concentrations ranging over 7-logs in magnitude. And, although sensitive immunoassays could be developed, the number of antibodies useful for that purpose is estimated to be around 10,000, or roughly 10% of the proteome. Furthermore, there is currently a limit to multiplexing using the sandwich immunoassay format. It is estimated that the number of simultaneous immunoassays that could be performed is well under 100 owing to problems with nonspecific interactions because of various levels of primary and secondary antibody–antibody interactions. So, the production of a human proteome chip based on antibodies is not likely for some time, or perhaps not at all. A better use of protein microarrays is in the development of small diagnostic panels and biomarker discovery platforms. Current roles for protein microarrays include adaptation of the sandwich immunoassay as a micro-ELISA in which capture antibodies are immobilized to the substrate, analyte captured, and its presence detected by the use of a secondary labeled antibody (Figure 1.2). This is referred to as the "forward" array format. Another area of interest is the development of the "reverse" array format, in which cell lysates are immobilized in an array and then interrogated with analyte-specific labeled antibodies. This is useful in determination of autoimmune diseases.

Beyond genomic and proteomic analyses are even higher-order biomolecular interactions that we are now just beginning to understand. Thus, we have included work on carbohydrate and lectin arrays. These will be important in order to examine the complex roles for glycoproteins, receptor proteins, and associated cell surface interactions.

Book Overview

The use of microarrays as a tool in analyzing genetic variation and gene expression is well documented. For the most part, this is a mature technology, although one could argue that because of the complexity and vast amounts of the data received, the task

Protein Microarrays are currently used in two common modes: micro-ELISA and differential display. These are antibody (probe) array formats in which purified antibodies, e.g., anti-cytokine, are printed onto the substrates. Micro-ELISA: Sample is incubated (containing antigens) and a labeled secondary antibody to antigen is added to form a sandwich assay for detection. Differential display: Control sample protein antigen labeled with Cy3, test sample with Cy5. Samples are mixed and applied to antibody array. Reverse (Antigen-printed) arrays are used to detect antibodies in a sample, e.g., autoimmune disease.

- Current Roles for Protein Arrays
  - Cytokine micro-ELISA
  - Autoimmune/Inflammation
  - Differential Display, e.g., Ciphergen

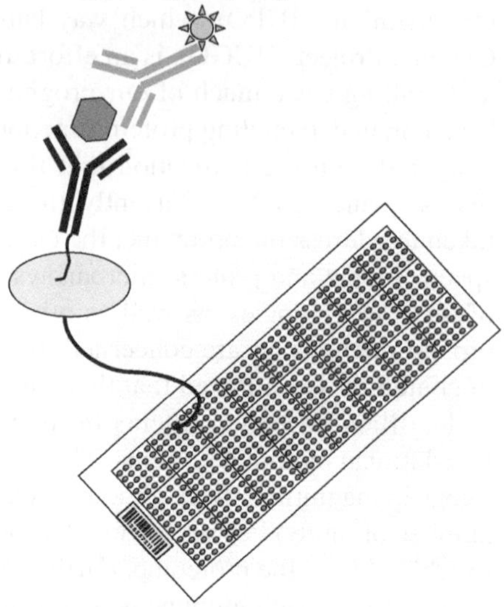

Figure 1.2
Protein microarrays. (Adapted from Kodadek, T., 2001.)

of analysis remains daunting. As far as adopting the microarray format in other areas such as, for example, proteomics or systems biology applications, we are very much at the beginning of its deployment. And, while our intent here is to provide you with a choice of "best in practice" methodologies to aid in achieving the highest-quality data, the reader must keep in mind that the microarray field continues to evolve. Misunderstandings in the past have occurred in their use and the interpretation of results. I am reminded of the over-used adage "garbage in, garbage out." In particular, do not forget the importance of the biological sample, for what you apply to the microarray can leave us all at times in awe of exquisite false-colored images of meaningless data. Where appropriate we will attempt to point out these "errors of our ways" and how best to apply microarrays, as well as gain a greater understanding of their strengths and limitations.

This book is intended to serve as a guide to setting up a successful microarray experiment. It is based on a collection of published methods and protocols, as well as the experiences of the author and collaborators. While we have attempted to provide a good set of tools, the reader should understand how they work and ascertain whether or not they can be appropriately applied in their work.

So, now that you realize this is not a recipe book, let us move on. The book is organized as a roadmap; that is, there is a starting point (sample preparation) and a destination (data analysis). Along the way, you pick up what you need to get to the destination. For example, if you intend to analyze proteins, start in the Protein chapter with protein isolation methods, move on to choosing, for example, protocols to prepare antibody arrays, and finish up with selecting a data analysis process. You will also find "suggested" routes provided for you. Here again, make sure that these make sense to you before starting out.

Roadmaps

These will provide you with an overview of, and orientation to, the various protocols. In all cases, the microarray roadmap is divided into two areas: (1) sample preparation and (2) microarray preparation, followed by assay development. These roadmaps give an outline of the book's coverage: Nucleic Acid Microarrays (Figure 1.3), Protein Microarrays (Figure 1.4), Carbohydrate Microarrays (Figure 1.5), and Lectin Microarrays (Figure 1.6). For example, The Nucleic Acid Microarray Roadmap (Figure 1.1) outlines the process of tissue homogenization, and then references a protocol, Protocol 2.1, that provides details on homogenizing tissues. If alternative or additional protocols are available, these are listed. For example, the extraction of nucleic acids is covered in Protocol 2.2, as well as Protocol 10.6.1. Each protocol is numbered to correspond to the chapter in which it is found. In the above case, Protocol 2.2 is found in Chapter 2 (Nucleic Acid Sample Preparation), whereas

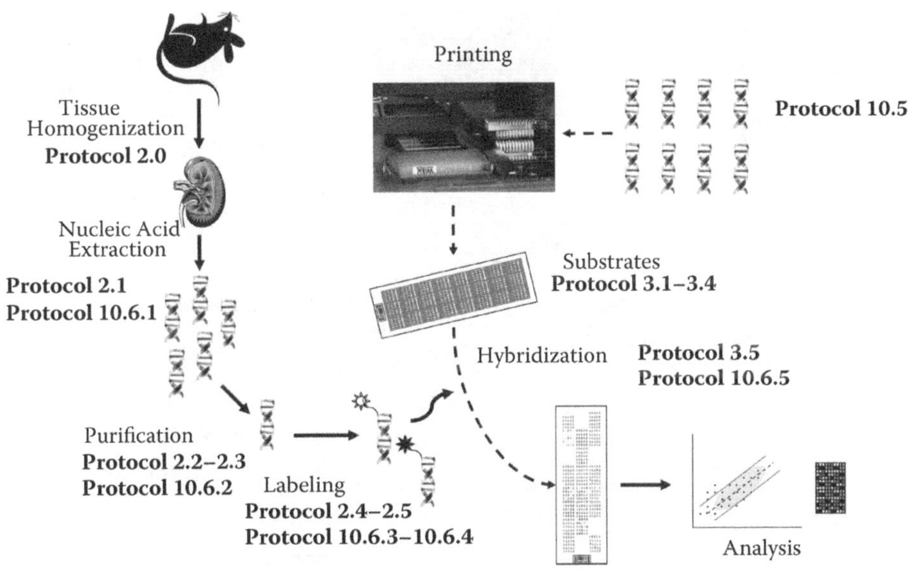

Figure 1.3
Nucleic acid microarray roadmap.

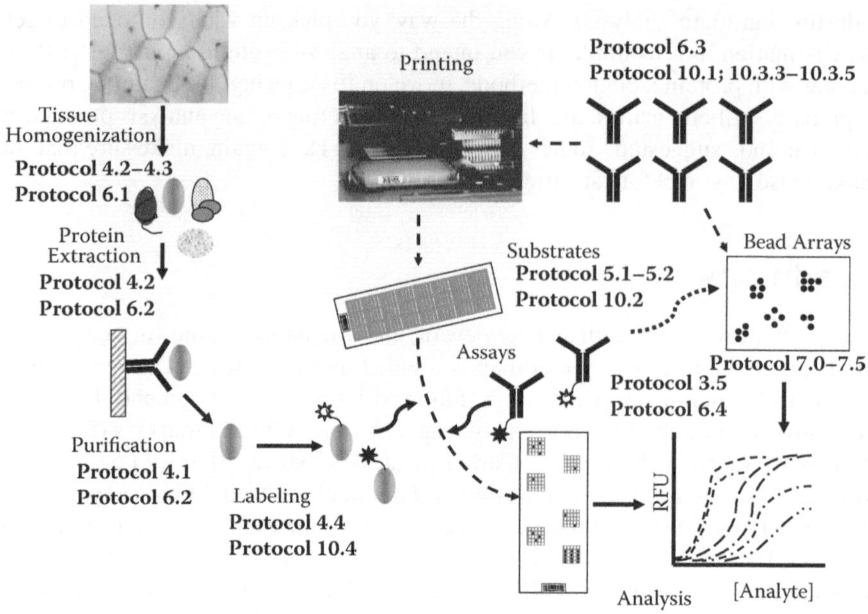

Figure 1.4
Protein microarray roadmap.

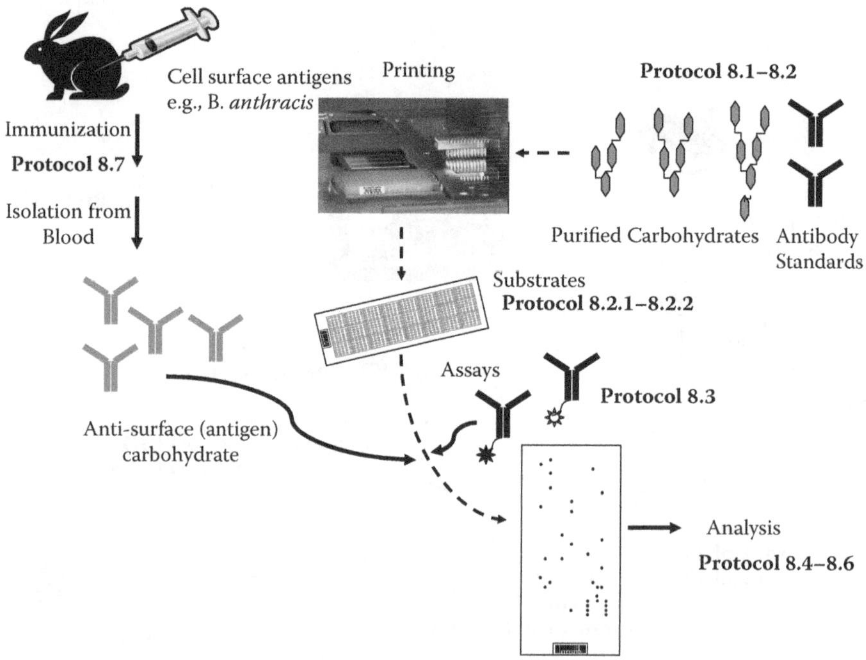

Figure 1.5
Carbohydrate microarray roadmap.

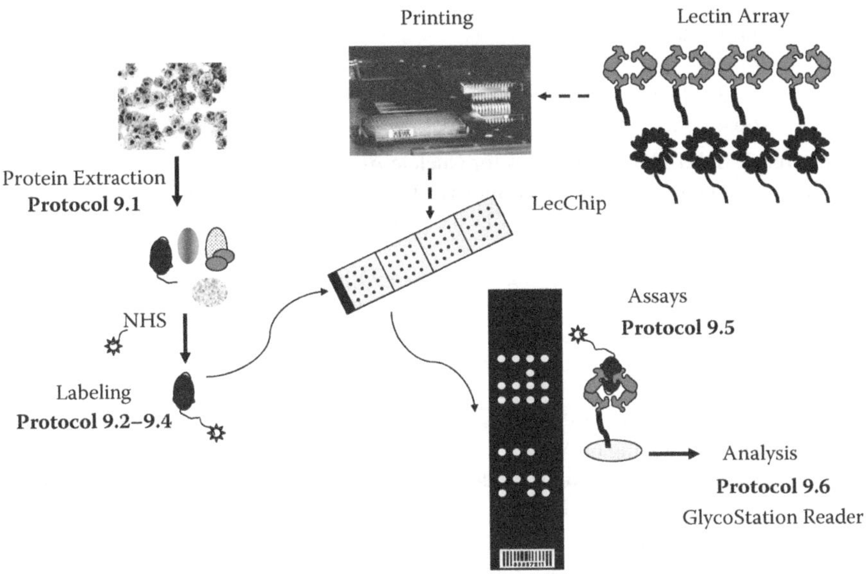

Figure 1.6
Lectin microarray roadmap.

Protocol 10.6.1 is found in Chapter 10 (Printing Methods). The complete list of chapters and corresponding protocols follows:

Protocol Format

Each of the protocols is arranged whenever possible using the format described later. This allows the reader to move quickly through each section. The intent is to provide a precise step-by-step process that will be easy to follow. Additional details regarding the protocols are found in chapter sections titled "How It Works" and "Troubleshooting Guide."

Each chapter is structured as follows:

The "Introduction" provides a general introduction to the topic, including recent advances. It cites important primary literature sources.

The "List of Protocols" contains a list of protocols incorporated into the chapter, for example,

Protocol Number	Name
2.1	Tissue Homogenization
2.2	Extraction of Nucleic Acids
2.3	RNA Purification
2.4	Electrophoresis of Nucleic Acids
2.5	Labeling of Nucleic Acid Targets
2.6	Storage of Nucleic Acids

Each protocol within the chapter is intended to provide a simple, succinctly described, and straightforward means of assistance to the user.

The "Title of Protocol" gives the protocol name.

"How It Works" consists of a short paragraph on the underlying principles.

"Required Materials" lists what materials, reagents, and equipment are required.

"Reagent Preparation" describes the preparation of major or special reagents.

The "Step-by-Step Protocol" presents a simple protocol enabling the user to move quickly through the procedure. It consists of step-by-step, single-line entries that are consecutively numbered. It does not include any notes or explanation. These are incorporated into the troubleshooting guide.

The "Key References" lists important primary literature references that best describe the protocol.

The "Troubleshooting Guide" describes any special notes, alternatives, and especially any pitfalls to be aware of and how to avoid them.

References

Freeman, W. M., (2000). Fundamentals of DNA hybridization arrays for gene expression analysis. *BioTechniques*, 29(5): 1042–1055.

Kodadek, T. (2001). Protein microarrays: Prospects and problems. *Chemistry & Biology*, 8(2): 105–115.

Matson, R. S. (2005) *Applying Genomic and Proteomic Microarray Technology in Drug Discovery*. Boca Raton, FL: Taylor & Francis.

Chapter 2

Nucleic Acid Sample Preparation

Robert S. Matson

Contents

Introduction

Extraction

The extraction of nucleic acids of high quality from samples is most important. However, different approaches are required depending on the sample source (e.g., mammalian or bacterial) as well as the kind of nucleic acid (e.g., genomic DNA or RNA) to be isolated. We will first examine the most widely accepted protocols for extracting DNA.

In most instances, isolation of a biomolecule from a crude extract relies on the physical-chemical properties of that molecule which distinguish them from other kinds of molecules found in the extract. For DNA, we exploit charge density and hydrophobicity. A strand of DNA can be regarded as a large anion or polyanion owing to the presence of the phosphate backbone, which imparts to the macromolecule a net negative charge. As a result, DNA will bind to a positively (cationic) charged material. In addition, the nucleic acid bases create hydrophobic batches within the strand. Based on these two properties (surface charge and hydrophobicity), DNA can be isolated onto a positively charged solid phase.

Purification

Target purification of the extracted nucleic acid is sometimes necessary for two reasons: first, as an enrichment process to increase the effective concentration and second, in order to remove contaminating species that may interfere with subsequent processing steps such as those prior to amplification. This may be simply the removal of proteins or cofactors containing metals that will inhibit PCR amplification. For example, if the sample is blood, then hemolysis can lead to the release of heme, which should be removed prior to PCR. In other instances, it may also be desirable to use a specific molecular species, for example, using mRNA in place of total genomic RNA from a sample. The purity and quantity of the target will have an impact on the extent of labeling and labeling strategy to employ.

Electrophoretic Analysis

It is prudent to know the state of the extracted sample prior to labeling and application to the microarray. The primary means of determining the character of the target in a sample is to perform electrophoresis. Agarose gel electrophoresis is the easiest and most informative method to determine whether or not the extracted DNA or RNA is suitable for further study. Essentially, you will obtain a molecular size profile of the nucleic acids in the sample. Both genomic DNA and RNA upon visualization (with, for example, ethidium bromide) have distinguishable banding patterns. If there is a highly smeared appearance (little banding observed) over the length of the gel, then degradation (fragments) most likely has occurred and the target quality in the sample is questionable. In the case of samples derived from an amplification process (e.g., RT-PCR) that results in increased abundance of a particular molecular weight species, these bands are predictable and should appear as distinct bands. The absence of smearing should be evident. In all cases, when running electrophoretic gels, it is advisable to include molecular weight markers or size standards. These will help you identify the relative size range of nucleic acids in your sample. A second advantage of running such markers is to verify whether or not the electrophoresis was properly set up and devoid of annoying artifacts.

In some instances, it may be desirable to examine the sample at a higher resolution than that afforded by agarose gels. A number of useful electrophoretic methods are available to do so. First, cross-linked polyacrylamide gels are available that provide various molecular sizing ranges. These gels can be prepared for use at a specified homogeneous composition commonly identified by the percentage (%) of cross-linker used, for example, 4%, 12%, and 20% gel. These will resolve nucleic acids of a particular molecular weight. The lower the percentage of cross-linker used, the greater the extent of migration of larger nucleic acid fragments into the gel, while small fragments migrate faster and appear earlier or near the leading edge of the dye front. Generally, the progress of the electrophoresis is monitored by watching colored "running" dyes (added in with the applied sample) migrate down the gel. There is usually a fast-running or frontal dye, and a slower-running or trailing dye incorporated into the sample. The nucleic acids migrate sandwiched between the two dye fronts. Once the frontal dye reaches the lower third of the gel, the electrophoresis is halted by turning off the power to the electrode, and the gel removed and stained to reveal the nucleic acid banding profile. In some cases, it may be useful to employ a gradient gel, for example, from 4% to 20% cross-linker. These gels allow one to examine both higher-molecular-weight nucleic acid species, as well as very low-molecular-weight species. When performing primer-based amplifications, gradient gels permit visualization not only of the amplicon target but also the primers. This permits verification of the efficiency of the primer-based amplification with sufficient resolution to observe primer-dimer formation, as well as any degradation or concatenation processes that may be occurring. Other electrophoretic methods used include capillary flow electrophoresis. Here, a glass capillary is filled with gel, and migration of the nucleic acid species monitored in real time using specialized optics.

Labeling Strategies

Microarray techniques are largely based on the reverse dot blot. That is, the probes are assembled as arrays on the microarray substrate, while the target is labeled and then applied to the probe array. It is important to understand what levels of sensitivity and specificity are required for the analysis prior to planning the labeling strategy. Two approaches are possible to meet robustness: mass amplification where label is incorporated during the amplification; or postlabeling of the targets. Which of these approaches is suitable for analysis depends on what is to be measured. For example, if the goal is to detect the presence or absence of a genetic mutation (e.g., SNP polymorphism), then mass amplification using end-labeled primers is a reasonable approach. Here, what is important is whether or not the SNP is present. If, however, the experiment is to measure differences in gene expression levels, then a nonlinear or exponential mass amplification process may not be appropriate because the relative levels of individual genes may vary depending on their respective abundance and the amplification process. Low-abundance genes may be amplified to a greater extent than the more abundant species in the sample.

Storage

One must use extreme caution when storing target samples. Whether purified target or extracted targets are to be stored, the presence of nucleases must be anticipated

Beware! RNase Is Everywhere!

RNA isolation:

"The quality of the RNA is the single most important determinant of a successful GeneChip analysis assay. Particularly, differential degradation of RNA can lead to erroneous conclusions about both the relative and absolute mRNA levels in the specimens."

Figure 2.1
RNase. (From http://keck.med.yale.edu/affymetrix/analysis.htm)

and the storage media prepared with nuclease inhibitors. RNase is particularly troublesome because this nuclease is not only ubiquitous but also very stable under a variety of harsh conditions, including autoclaving (Figure 2.1). Although chelating agents (e.g., EDTA) are useful in preventing nuclease activity, prolonged storage may lead to removal of magnesium ion from the nucleic acid, thereby destabilizing the helix and causing fragmentation to occur.

List of Protocols

Protocol Number	Name
2.1	Tissue Homogenization
2.2	Extraction of Nucleic Acids
2.3	RNA Purification
2.4	Electrophoresis of Nucleic Acids
2.5	Labeling of Nucleic Acid Targets
2.6	Storage of Nucleic Acids

Protocol 2.1: Tissue Homogenization

How It Works

The adage "garbage in, garbage out" is especially applicable to the isolation of macromolecules from tissues and cells. In particular, RNA is highly susceptible to degradation by RNases, and poor-quality RNA has been the ruin of many microarray-based experiments. The problem with RNases is that they are everywhere (including your hand gloves) and extremely difficult to destroy. The best approach is to first inhibit the enzyme during extraction and then hopefully, as a consequence of the RNA isolation process, leave it behind.

Many protocols to prepare good-quality RNA are available. The problem is selecting which of these to use for a particular tissue or cell type. All protocols must

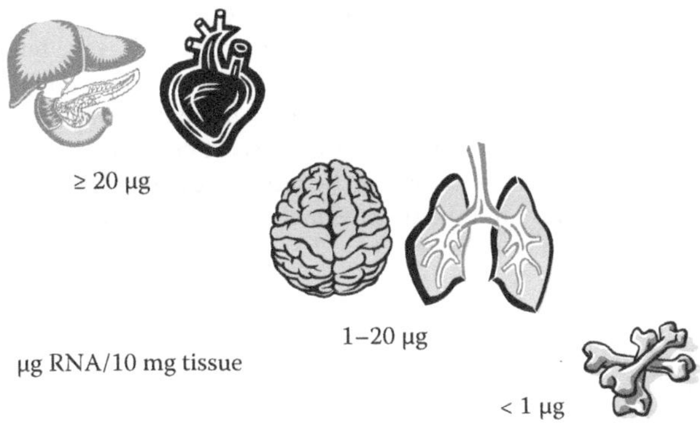

Figure 2.2
Total RNA recovered from various tissues.

start with some means of rupturing the cell and releasing the RNA. The most common method is homogenization. This consists essentially of applying physical forces to the tissue to open the cell membrane: shear, rip, tear, smash, pulverize, explode, implode, freeze-fracture, and dissolve; and, of course, all on ice to preserve that which we wish to isolate for further study. The following protocols are best regarded as examples because variations in tissue and cell types most assuredly will require the individual to perform further optimizations (Figure 2.2).

Required Materials

Homogenizer Types: Select type based upon tissue and preparation scale

Dounce: Usually consists of glass conical tube to which a glass pestle precisely fits. Tissue is sheared between the pestle and wall of the tube. It is recommended to have several sizes with different clearances between tube and pestle rod. This allows for stepwise shearing to produce complete homogenization.

Potter-Elvehjem: Comprises a glass conical tube to which a stainless steel pestle with a Teflon-coated bottom is precisely fitted. Unit can be used for manual grinding or the shaft of the pestle inserted into an overhead motor for power-driven shearing.

Waring blender: Device consists of a jar (glass or stainless steel) that sits on a motorized base having stainless steel rotor blades that face up into the seated jar. Tissue is added through the jar lid along with homogenizing buffer. The tissue is subjected to the blade shear and chopped or blended depending on the rotor speed that is applied.

Polytron (rotor-stator): In this case, the tissue is liquefied (like a milkshake) as it is taken up into the center portion of a motorized mixing tip. There it is subjected to high shear forces and rigorously mixed by the rotor-stator and flowed past the stator blades (teeth) to complete homogenization.

Sonicator: Produces ultrasonic waves that cause cells to burst open by cavitation. The vibrating tip (probe) causes the formation of microbubbles in the media. These implode, producing shock waves that disintegrate the cell wall. The tip also causes microstreaming of these bubbles, sending them colliding with the cell surface and creating a high shear force. The net result from cavitation and microstreaming is rapid bursting of the cell contents into the media. The downside of this approach is the buildup of heat from the process, which can cause denaturation of macromolecules, and the unwanted generation of free radicals and hydrogen peroxide. You should always sonicate your sample cool, preferably within an ice bath. A series of short or pulsed sonications will reduce the buildup of heat. Most sonicators can be tuned to disrupt different tissues.

Bead mill: Cells or tissues are mixed with small glass, stainless steel, or ceramic beads and rigorously mixed. The beads crush or shear the cell wall. The beads can be removed by centrifugation.

Mortar and pestle: This manual method of grinding tissue works well with liquid nitrogen frozen tissue or cell pellets.

French press: Tissues are pressed through a small orifice that creates high shear forces and breaks the cells open.

Parr bomb: Cells are subjected to rapid nitrogen decompression ("the bends"), causing the cell wall to burst open. First, cells and tissues are suspended in a medium and then placed inside the pressure cylinder. The cylinder is cooled, sealed, and pressurized with nitrogen gas. Then, the pressure is suddenly released, at which point the cells burst.

Needle shearing: If the homogenate remains too viscous, it can be passed back and forth through a fine-gauge needle. Use a sterile 18 G needle and syringe. Pass the homogenate through it several times, then replace with a 20 to 22 G needle. Repeat the process until the pressure on the needle is reduced.

Other Materials

Glassware: All glassware must be freshly cleaned. Start by rigorously cleaning with a detergent, followed immediately by a thorough rinse (e.g., 3–6 times) in deionized water. Place the cleaned glassware in a drying oven overnight at 100°C. Place aluminum foil over the tops of containers.

RNase-/DNase-free tubes, pipette tips, and other disposables: Use these whenever possible.

Gloves: At all times wear disposable gloves. It is recommended to double gloves if possible and change them periodically to avoid carryover contamination.

Reagents and Buffers

RNase Zap (Ambion) or RNase AWAY(Invitrogen): Use to decontaminate glass, plasticware, benches, and equipment of RNase.

RNAlater (Ambion): Use to stabilize RNA in freshly prepared tissue and cells prior to extraction.

Ethanol: 70% (v/v) in deionized water for decontamination.

Reagent Preparation

Diethylpyrocarbonate (DEPC): Used to chemically inhibit RNase by modification of histine residues on the enzyme. Prepare deionized, distilled water by adding 0.1% (v/v) DEPC for at least 1 h to overnight (better) at 37°C. Autoclave the treated water for 15 min to destroy residual chemical. This water is now RNase-free. Note: DEPC is amine reactive, so it should not be used directly with amine-containing buffers such as Tris or HEPES because these would consume the reagent and reduce the level of RNase inactivation.

Step-by-Step Protocols: 2.1.1–2.1.3

Start with ~10 mg to 50 mg animal tissue. Note: Many microarray experiments now require less than 10 µg total RNA for labeling and hybridization. Using at least 10 mg tissue will permit isolation of 10–30 µg total RNA for that purpose (Figure 2.3).

Required Materials

Animal organ, tissue, or cells
Ethanol (70%)
Scalpels (sterile)
Saline solution (0.9% sodium chloride; sterile)
Weighing paper or boats
Balance (0.00 g)
RNAlater
Dry ice
Liquid nitrogen (LN_2)
Cryovials with screwcap

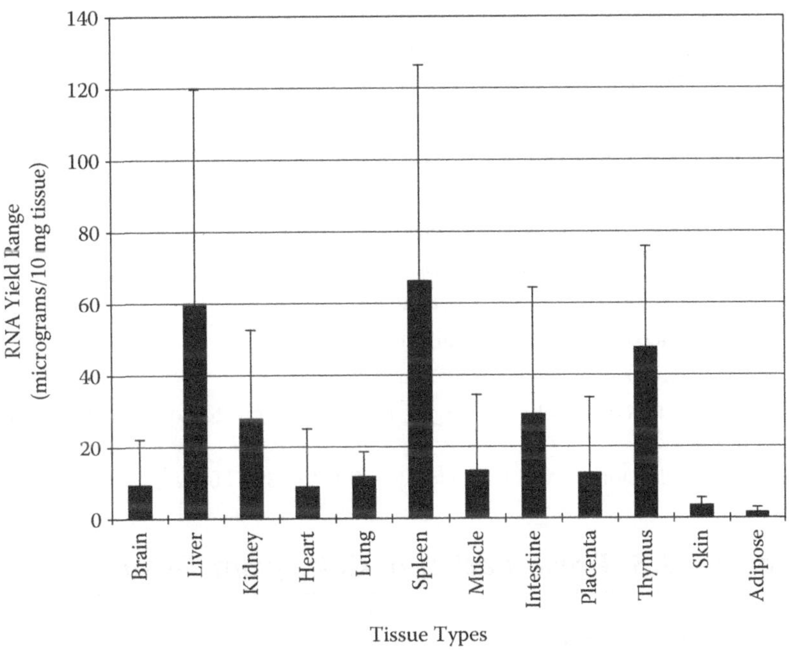

Figure 2.3
RNA yield ranges for different tissue types.

Protocol 2.1.1: Tissue Preservation Using RNAlater

Step-by-Step Protocol

1. Kill a rat, and clean its underside with 70% ethanol.
2. Make an incision in the abdomen, and remove the liver using a sterile scalpel. Avoid cutting through blood vessels or gallbladder.
3. Rinse the liver tissue in ice-cold sterile saline, then blot dry on Kimwipes.
4. Use a new sterile scalpel to excise approximately 2–3 mm cubes of tissue. Special note: ~3 mm cube of animal tissue weighs ~30 mg.
5. Place each tissue cube on weighing paper (or boat), and obtain the cube's mass using a pan balance.
6. If necessary, further dice the tissue cube to adjust to a mass within the desired range. However, ensure that you complete above steps within 2–3 min of the excision (step 4, above).
7. Immediately submerge the tissue cube in the appropriate volume of RNAlater. The volume is calculated as follows: Tissue cube mass (mg) × 10 μL/mg = total volume RNAlater.
8. The stabilized RNA tissue cube may be stored for ~1 month at 2–8°C.
9. Remove the tissue cube from the RNAlater solution.
10. Transfer the tissue cube into a conical tube previously placed on ice for homogenization.
11. Proceed to Extraction of Nucleic Acids (Protocol 2.2).

Protocol 2.1.2: Tissue Preservation Using Liquid Nitrogen (LN_2) Snap-Freezing

Step-by-Step Protocol

1. Place dry ice in a Styrofoam™ container.
2. Cool down a clean, dry plastic beaker in the container.
3. Pour LN_2 from a Dewar into the beaker.
4. Follow steps 1–6 from Protocol 2.1.1 (above).
5. Place the diced tissue in a cryovial, and screw on cap.
6. Immediately drop vial into beaker containing LN_2.
7. Maintain vial completely submerged in LN_2 until transfer to a LN_2 Dewar for long-term storage.
8. Alternatively, store cryovial samples in –70°C to –80°C freezer.

Protocol 2.1.3: Preservation of Cells from Tissue Culture

Step-by-Step Protocol

All steps should be conducted with aseptic technique. See Figure 2.4 regarding expected total RNA yields from cells.

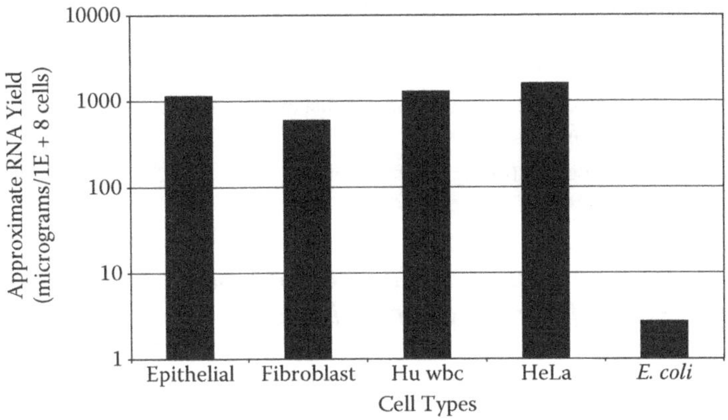

Figure 2.4
Total RNA yield from various cells.

1. Initial collection of cells:
   a. For nonadhering cells in suspension culture, remove from media by centrifugation. Most cells can be collected as a pellet by centrifugation at low speed using either a swinging bucket or fixed-angle rotor. Centrifuge (4°C) at ~200–500× g, 5–10 min depending on the cell type and rotor.
   b. For adhering cells that form a monolayer in T-flasks, it is first necessary to detach these from the plastic's surface prior to collection by centrifugation. This can be accomplished by either physically scraping off the cells using a cell scraper or by the biochemical process of trypsinization (see Protocol 2.1.4).
2. Place the centrifuge tube or bottle on ice.
3. Observe the condition of the cell pellet. It should not appear to be loosely packed or sloughing off. Increase centrifugation time or speed if this occurs. However, avoid excessive g-force, because cell rupture can occur.
4. Remove the supernatant while taking care not to dislodge the pellet. This can be accomplished either by aspirating off the liquid or by carefully pouring off the liquid from the centrifuge tube. In either case, avoid any back flow of the fluid into the centrifuge tube. Always aspirate or pour from the sidewall furthest away from the pellet.
5. Wash the cell pellet to remove residual media by first resuspending the pellet in ice-cold PBS buffer. This can be accomplished by several different means, depending on the nature of the pellet, tube geometry, etc.
   a. **Rubber policeman:** Using the rubber policeman attached to a glass rod, gently rub the pellet to dislodge. Avoid dislodging the intact pellet into the buffer as this may make it more difficult to resuspend. Rather, stepwise add a small volume of buffer and resuspend a portion of the pellet. Repeat this process until the pellet is completely dissolved, then add additional buffer and gently mix to ensure uniform resuspension.
   b. **Pipette aspiration:** Partially fill a glass Pasteur pipette with buffer. Gently rinse the pellet with the pipette volume. Aspirate the rinse, and pass it over the pellet until it has largely dissolved into the buffer. Add additional amounts of buffer and aspirate-dispense back and forth in the pipette to produce the cell suspension.

c. **Vortex:** Add a small volume of buffer to the tube, and wet out the pellet. Gently vortex, avoiding splatter and excessive frothing. Add additional buffer and repeat process until pellet is in suspension.

6. Centrifuge the PBS cell suspension, and discard the supernatant. Repeat steps 1–4 for a total of three times.
7. Snap-freeze the pellet in LN_2 and store at –70°C to –80°C until ready for extraction. Alternatively, cells may be prepared as a suspension in 5–10 mL RNAlater.

Protocol 2.1.4: Trypsinization to Remove Adherent Cells from Culture Flask

How It Works

Adherent cells attach to plastic and glass surfaces by the excretion of cell surface proteins that anchor the cells. Trypsin is a proteolytic enzyme that can effectively digest the anchoring proteins. However, the enzyme is inhibited by certain metal ions (divalent cations: calcium, magnesium) found in the growth media. EDTA is added to chelate these and thereby preserve the enzymatic activity. Because trypsin would continue to digest away cellular proteins, it is important to allow only a limited digest or just enough to detach the cells and not injure them. So, serum-based media containing endogenous inhibitors is added back after detachment in order to arrest further trypsin digestion.

Required Materials

Cultured cells
Cell culture media
Trypsin
EDTA
DDI water (sterile)
Incubator (37°C)
Dulbecco's PBS (calcium, magnesium free) or PBS (–), sterile
Microscope (phase contrast, inverted-type preferred)
Centrifuge
Centrifuge tubes (sterile)
Pipettes (sterile)
Hemacytometer (cell counts)

Reagent Preparation

Prepare fresh 0.25% trypsin-EDTA solution: add 0.25 g trypsin and 0.25 g EDTA to a final volume of 100 mL of DDI water. Incubate at 37°C to prewarm prior to addition to cells in flask.

Step-by-Step Protocol

All steps should be conducted with aseptic technique.

1. Aspirate or pour off culture media from culture flask.
2. Add back PBS(–) buffer to rinse cells free of residual media. Discard rinse. Repeat rinse one more time.

3. Add prewarmed trypsin-EDTA to culture flask.
4. Incubate at 37°C for 5–10 min.
5. Upon detachment, add excess fresh culture media to flask.
6. Transfer cell suspension to centrifuge tube, and centrifuge at ~200 to 500 × g for 5–10 min to pellet cells.
7. Process according to Protocol 2.1.3, steps 2–6.
8. Remove an aliquot, and determine cell count (# cells/mL) after first rinse.
9. Centrifuge to pellet cells.
10. Snap-freeze the pellet in LN_2 and store at −70°C to −80°C until ready for extraction. Alternatively, cells may be prepared as a suspension in 5–10 mL RNAlater.

Key References

Foley, J. F. et al. (2006). Optimal sampling of rat liver tissue for toxicogenomic studies. *Toxicologic Pathology*, 34(6): 795–801.
RNAlater Handbook, Qiagen Corporation, July 2006.

Troubleshooting Guide

Note 1: Trypsin Addition Select a volume to add that just covers the bottom of the flask.
Note 2: Cell Detachment Observe cells in flask under microscope. Cells should be detached (floating) or rounded if loosely adsorbed to surface.
Note 3: Cell Counts It will be necessary later to adjust to ~10^6 cells/mL for extraction purposes.

Protocol 2.2: Extraction of Nucleic Acids

Once tissue has been homogenized, the RNA must be quickly removed to avoid degradation by RNases. The Boom method (Protocols 2.2.1 to 2.2.3) relies upon adsorption of RNA to silica particles, while the Trizol method (Protocols 2.2.4 and 2.2.5) removes contaminents and isolates RNA by solution-phase partioning.

Protocol 2.2.1: Boom Method for mRNA Isolation

How It Works

The principle of the method is based on the use of silica particles that will bind nucleic acids and some proteins under high-salt conditions. Guanadine thiocyanate (or guanadine hydrochloride) is used to lyse cells, inhibit RNAase activity, and promote binding to the silica (Figure 2.5). Diatoms are microscopic aquatic organisms of a uniform size that secret silicaceous acid as a protective biological barrier. Their skeletons are collected in diatomaceous earth and serve as the silica particle source.

Required Materials

Silica particles: diatomaceous earth (Celite)
Centrifuge: 12,000× g-force
Guanidine thiocyanate (GuSCN)

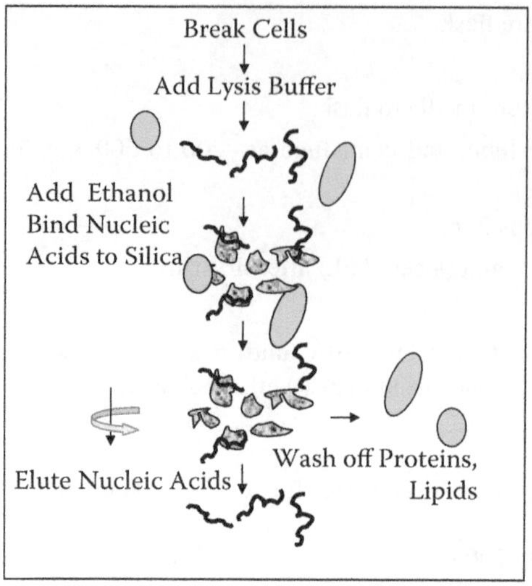

- The Boom Method
  - Lysis buffer
    - Chaotropic
    - Inactivates RNase
    - 4.7 M GuSCN with Tris-HCl, EDTA, Triton X-100
  - Additives
    - Add Proteinase K to digest protein
    - Add DNase to digest dsDNA
  - Silica
    - SiO_2 particles; glass; diatoms
    - Binds mostly nucleic acids, some protein, lipid
  - Pure RNA
    - A260/A280 = 1.9–2.0
    - 1 OD260 = 40 μg/mL RNA
    - A260/A230 < 2.0 GuSCN present

Figure 2.5
The Boom method. (Adapted from Boom et al., 1990.)

Tris
EDTA
Ethanol (70%)
Acetone
Triton X-100
DEPC-treated water

Reagent Preparation

Wash buffer: GuSCN, 120 g dissolved in 100 mL 0.1 M Tris, pH 6.4. Make an extra amount for preparing the lysis buffer (below).
Lysis buffer: Add to 100 mL wash buffer, 22 mL 0.2 M EDTA, pH 8.0 + 2.6 g Triton X-100
Elution buffer: Tris-EDTA, pH 8.0 (10 mM Tris-HCl + 1 mM EDTA)

Step-by-Step Protocol

1. Combine 900 μL lysis buffer with 40 μL of diatomaceous earth in an Eppendorf tube.
2. Vortex.
3. Add 50 μL sample to tube.
4. Vortex.
5. Let stand for 10 min.
6. Vortex.
7. Centrifuge.
8. Suction off supernatant and discard.
9. Resuspend pellet in wash buffer.
10. Wash pellet 2 times with 1 mL of ethanol (70% v/v) resuspended by vortex and centrifugate to pellet.

11. Add acetone to the pellet, vortex, and recollect pellet by centrifugation.
12. Dry pellet on a heat block for 10 min at 56°C.
13. Resuspend pellet in elution buffer and incubate for 10 min at 56°C.
14. Vortex and centrifuge at 12,000× g for 2 min.
15. Recover the supernatant as the source of DNA + RNA.

Key Reference

Boom, R. et al. (1990). Rapid and simple method for purification of nucleic acids. US Patent 5,234,809. *Journal of Clinical Microbiology,* 28(3): 495–503.

Protocol 2.2.2: RNA Isolation Using Spin Columns

How It Works

Most spin column methods rely on membrane impregnated in some manner with silica or finely crushed glass particles (see Vogelstein and Gillespie, 1979) in place of diatomaceous earth. The lysate is applied to the spin column, the nucleic acids are bound to the silica gel, and the RNA recovered after removal of contaminating proteins and DNA (Figure 2.6). The advantages are simplicity of use and more consistency in extraction. The following is adapted from Invitrogen's (Carlsbad, CA) S.N.A.P. spin-column protocol. Similar protocols are available from other vendors. However, bear in mind that isolation kit buffers and other components will vary depending on kit design. Certain reagents used in these kits may be regarded as proprietary and are not disclosed to the user.

Required Materials

Spin columns with silica gel impregnated membrane
Centrifuge
Isopropanol (IPA)
Guandine thiocynante (GuSCN)
Guanidine HCl (Gu-HCl)
Triton X-100
Tris
$MgCl_2$
$CaCl_2$
DNase I (2U/μL)
DEPC-treated water

Reagent Preparation

Lysis buffer: Several versions are available, for example:

1. Chirgwin method: 4 M GuSCN containing 25 mM sodium citrate, pH 7.0, then add ß-mercaptoethanol (1% final conc., 10 μL per mL buffer) and sarcosyl (0.5% final concentration) just prior to use.
2. Chomczynski and Sacchi method:
   Trizol buffer
   Binding buffer: 7 M Gu-HCl containing 2% Triton X-100

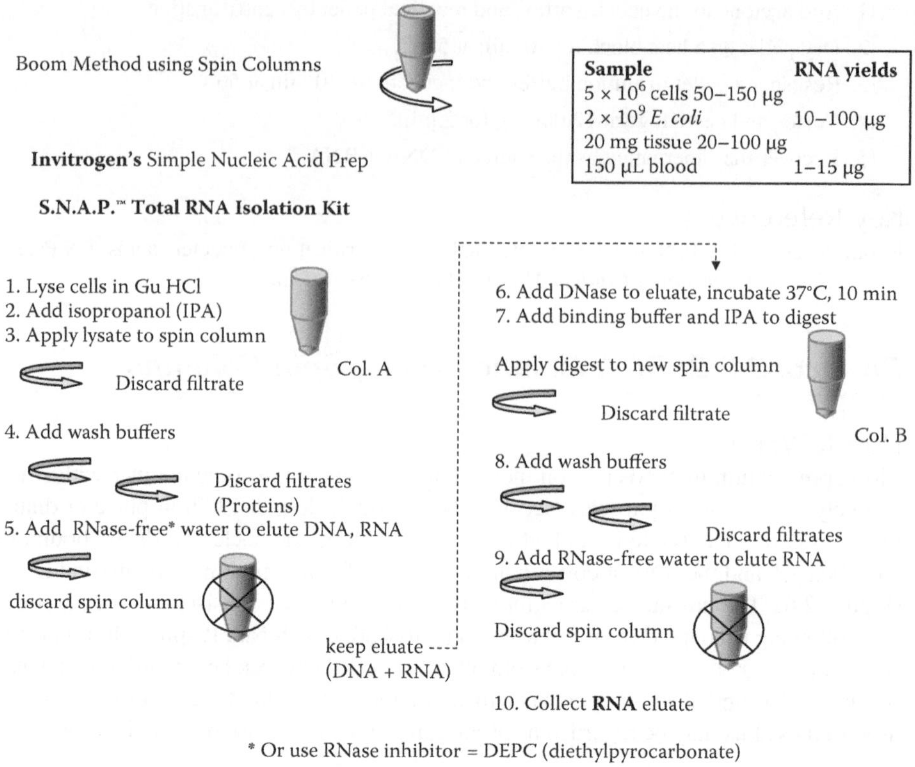

Figure 2.6
Boom method using spin columns. (Adapted from Cat# K1950 User Manual. http://www.Invitrogen.com.)

Wash buffer: 5.25 M Gu-HCl containing 1% Triton X-100
Elution buffer: RNase-free (DEPC-treated) water
DNase buffer: 0.4 M Tris Buffer, pH 8, containing 60 mM $MgCl_2$ and 20 mM $CaCl_2$

Step-by-Step Protocol

1. Lyse cells.
2. Add IPA.
3. Apply lysate to the spin column.
4. Spin.
5. Discard filtrate.
6. Add wash buffer.
7. Spin.
8. Discard filtrate.
9. Add DEPC water.
10. Spin.

11. Recover eluate.
12. Discard spin column.
13. Add DNase to eluate, and incubate at 37°C for 10 min.
14. Add binding buffer and IPA to digest.
15. Apply digest to new spin column.
16. Spin.
17. Discard filtrate.
18. Add wash buffer.
19. Spin.
20. Discard filtrate.
21. Add RNase-free water for elution of RNA.
22. Spin, and discard spin column.
23. Recover RNA eluate.

Key References

Extraction of RNA

1. Chirgwin, J. M. et al. (1979). Isolation of biologically active ribonucleic acid from sources enriched in ribonuclease. *Biochemistry*, 18: 5294–5299.
2. Vogelstein, B. and Gillespie, D. (1979). Preparative and analytical purification of DNA from agarose. *Proceedings of the National Academy of Sciences USA*, 76(2): 615–619.
3. Chomczynski, P. and Sacchi, N. (1987). Single-step method of RNA isolation by guanidinum thiocyanate-phenol:chloroform extraction. *Analytical Biochemistry*, 162: 156–159.
4. Invitrogen Corporation (2001). S.N.A.P.™ Total RNA Isolation Kit User Manual, Version G. www.invitrogen.com
5. Invitrogen Corporation (2006). PureLink™ Micro-to-Midi Total RNA Purification System User Manual.
6. SuperArray Bioscience Corporation (2006). ArrayGrade™ Total RNA Isolation Kit User Manual.

Extraction of DNA

7. Borodina, T. A. et al. (2003). DNA purification on homemade silica spin-columns. *Analytical Biochemistry*, 321: 135–137.

Protocol 2.2.3: RNA Isolation Using Filtration Manifold

How It Works

In this case, the silica-impregnated membrane is housed in a filtration device. This allows for scale up in multiple sample processing. For example, a 96-well filter bottom plate with vacuum manifold permits the preparation of 96 RNA samples. The extraction and purification steps conveniently take place on the filter (Figure 2.7). A further advantage to this approach is that robotic liquid-handling system can be

Boom Method Using Spin Columns or Filtration Manifold

RNeasy MinElute Cleanup Kit
GuSCN Lysis Buffer
Silica gel impregnated membrane-based spin column
Binding capacity ≤ 45 µg RNA
Starting volume ≤ 200 µL
Elution volume 10–14 µL
Preparation time < 15 minutes

Tissue	RNA yield per 10 mg
Liver	15–80 µg
Heart	5–25 µg
Lung	5–15 µg

RNeasy 96 Universal Tissue Plate
QIAzol (phenol-Gu "salt"/chloroform) Lysis Buffer
Silica gel impregnated membrane-based spin column or vacuum filtration formats (96-well plate)
Binding capacity ≤ 100 µg RNA per well
Starting volume, 1 mL per well
Elution volume ~ 100 µL per well
Preparation time ~ 2 hours

Figure 2.7
Qiagen's RNeasy.

applied to integrate and automate the processing. The following abbreviated protocol serves as an example.

Required Materials

96-well flat bottom plates
96-well filter plates
96-well collection plates
Filtration vacuum manifold
Robotic liquid handling workstation (optional)
Lysis buffer (varies depending on cell or tissue)
Cultured cells (or homogenized tissue)
Ethanol, reagent-grade
Ethanol, 70%
Wash buffer: e.g., 0.1 M Tris, pH 8.8, 0.2 M NaCl in 80% (v/v) ethanol
DNase I (1–2 U/µL)
DNase buffer: e.g., 10 mM Tris-HCl, 2.5 mM $MgCl_2$, 0.5 mM $CaCl_2$, pH 7.6
Stop buffer: e.g., 20 mM EGTA, pH 8.0 in 60% (v/v) ethanol

Step-by-Step Protocol

1. Dispense cells into flat bottom plate.
2. Dispense lysis buffer into flat bottom plate.
3. Mix by pipetting.
4. Apply lysates from 96-well flat bottom plate to the 96-well filter plate (seated with collection plate in vacuum manifold).

5. Block off any unused wells of the filter plate with sealing tape.
6. Vacuum-filter lysate solution.
7. Vent filter plate.
8. Add wash buffer.
9. Vacuum-filter wash buffer.
10. Vent filter plate.
11. Discard collected filtrate (waste).
12. Replace collection plate.
13. Add DNase solution to filter plate wells.
14. Incubate for 10 min.
15. Add stop solution.
16. Vacuum-filter.
17. Add wash buffer.
18. Vacuum-filter until filter is dry.
19. Disassemble filtration device.
20. Remove collection plate and discard filtrate.
21. Replace collection plate and reassemble filter unit.
22. Add RNase-free water to elute RNA.
23. Vacuum-filter.
24. Recover RNA eluate in collection plate.

Key References

Cell Culture RNA

Shultz, S. (2007). Promega Applications Note 171.

Mahadevappa, M. and Warrington, J. A. (1999). A high-density probe array sample preparation method using 10- to 100-fold fewer cells. *Nature Biotech*, 17: 1134–1136.

Tissue RNA

Qiagen Corporation (2004). RNeasy 96 Universal Tissue Handbook.

Plasmid DNA

Itoh, M. et al. (1999). Automated filtration-based high-throughput plasmid preparation system. *Genome Research,* 9: 463–470.

Protocol 2.2.4: Trizol Method for RNA Isolation

How It Works

This method relies on solution-phase partitioning and extraction of nucleic acids, as well as proteins. Cells are lysed in GuSCN containing phenol. The addition of chloroform creates an aqueous phase portioned from an organic phase. The aqueous

Tissue	RNA yield per 10 mg
Liver	60–100 μg
Kidney	30–40 μg
Cells	**per 10^6**
Epithelial	8–15 μg
Fibroblast	5–7 μg

TRI Reagent

Invitrogen's TRIzol

TRIzol® Reagents are ready-to-use, mono-phasic solutions of phenol and guanidine isothiocyanate suitable for isolating total RNA, DNA, and proteins. Isolation procedures are based upon improvements to the single-step RNA isolation method developed by Chomczynski and Sacchi and are completed in less than one hour.

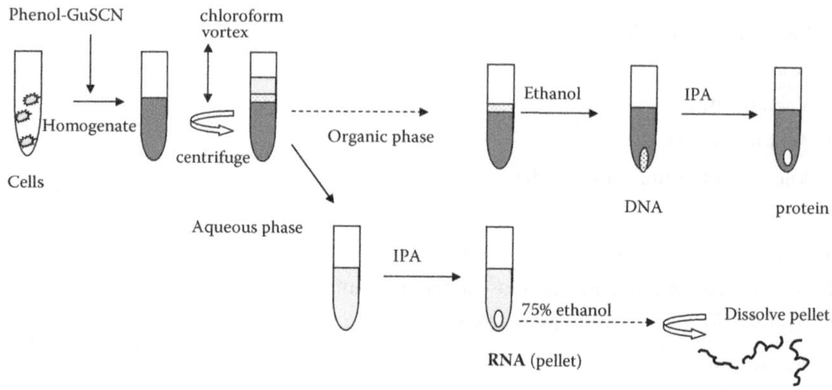

Figure 2.8
Trizol method for RNA isolation. (From Chomczynski, U.S. Patents 4,843,155 and 5,346994; Molecular Research Center, Inc., Cincinnati, OH)

phase contains the RNA, whereas DNA and proteins reside in the organic phase. Each of these can be isolated by alcohol precipitation (Figure 2.8).

Required Materials

TRIzol reagent (Invitrogen) or
TRI reagent (MRC) or
GuSCN–phenol extraction solution (see "Reagent Preparation," Note 1)
Chloroform
Ethanol
Isopropanol
Eppendorf tubes
Dounce tissue homogenizer (glass-Teflon)
Centrifuge
Pipettor

Reagent Preparation

Guanidinium–phenol extraction solution: Combine 468 mL GuSCN (4 M) with 495 mL phenol. Add 25 g sodium acetate and 1.8 mL 2-mercaptoethanol. Mix well. Adjust to pH 4 with acetate buffer.

Step-by-Step Protocol

1. Place 25 mg tissue into homogenizer, and add 800 μL extraction solution.
2. Form homogenate.
3. Transfer homogenate to Eppendorf tube.
4. Add 80 μL of chloroform to tube, and shake for 15 s to form suspension.
5. Place suspension on ice for 15 min.
6. Centrifuge at 12,000× g for 14 min at 4°C to separate aqueous (upper) from organic (lower) phase.
7. Transfer upper aqueous phase to new tube with pipette.
8. Add equal volume of isopropanol to tube and mix.
9. Place tube in –20°C freezer for 45 min.
10. Centrifuge at 12,000× g for 14 min at 4°C.
11. Remove supernatant, discard.
12. Observe white pellet (RNA) in tube.
13. Wash pellet by vortexing up into 800 μL of 75% ethanol–water.
14. Centrifugation at 12,000× g for 8 min at 4°C.
15. Repeat washing process (step 12).
16. Dry pellet under vacuum for 10 min.
17. Dissolve RNA in TE buffer.

Key References

Chomczynski, P. (1989). U.S. Patent 4,843,155.

TRI Reagent (1995). RNA, DNA, protein isolation reagent, Manufacturer's protocol. Molecular Research Center, Inc., Cincinnati, OH.

Troubleshooting Guide

Note 1: Extraction Reagent Trizol is the trade name held by Invitrogen but is commonly used to describe Chomczynski's reagent. Tri-reagent is the trade name held by Molecular Research Center for the extractant.

Protocol 2.2.5: Trizol Method for RNA Isolation Using Magnetic Beads

How It Works

This is a combination of methods (Trizol + Boom) aimed at higher-throughput applications that require 96-well microtiter plate formats. Cells are first extracted using the Trizol reagent in which RNA is separated out into the aqueous phase. The RNA is then bound to magnetic beads (silica) and processed essentially by the Boom method using a magnetic plate to capture, wash, and eventually elute pure RNA

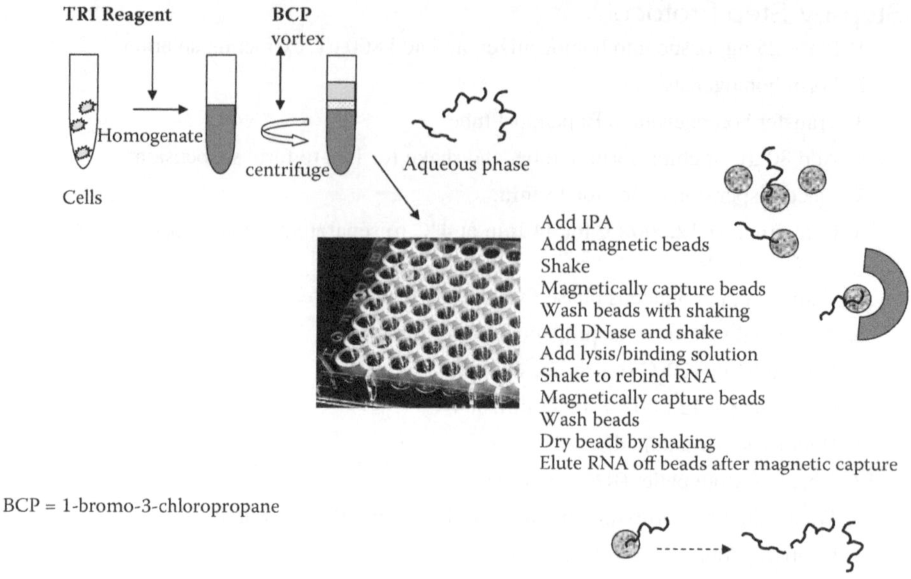

Figure 2.9
Trizol method using magnetic bead separation; Ambion's MagMax™-96 for microarrays kit.

from the beads. The Ambion protocol is described for producing multiple samples for microarrays (Figure 2.9).

Required Materials

TRI reagent (Ambion's reagent comprising phenol + GuSCN; store 4°C)
Ethanol
Isopropanol
Lysis/binding enhancer (Ambion; store –20°C)
TURBO DNase (Ambion; store –20°C)
Eppendorf tubes
96-well microtiter plate
96-well magnetic capture plate
BCP (1-bromo-3-chloropropane)
MagMAX beads (Ambion; store 4°C)
Dounce tissue homogenizer (glass-Teflon)
Orbital shaker
Pipettor (or automated liquid handling robotic workstation)

Reagent Preparation

Ambion lysis/binding buffer: Add 6 mL isopropanol to concentrate and mix well.
Ambion wash buffer I: Add 6 mL isopropanol to concentrate and mix well.
Ambion wash buffer II: Add 58 mL ethanol to concentrate and mix well.
Store these reagents at room temperature.

Enhancer solution: for 96 wells, mix 15.4 mL wash buffer I with entire contents (1.1 mL) of lysis/binding enhancer. Prepare just before use (step 14 below). TURBO DNase solution: for 96 wells, mix 5.3 mL MagMAX TURBO DNase buffer with entire contents (220 μL), and mix well. Prepare just prior to use (step 23 below).

Step-by-Step Protocol

1. Place 25 mg tissue into Eppendorf tube used for homogenation, and add 500 μL TRI reagent.
2. Form homogenate using homogenizer.
3. Incubate homogenate 5 min, room temperature.
4. Deliver 10 μL BCP to each well of a 96-well microtiter plate.
5. Add 100 μL of a homogenate sample into each well.
6. Place plate on orbital shaker, and shake vigorously for 1 min.
7. Deliver 50 μL IPA to each well.
8. Place plate on orbital shaker, and shake vigorously for 1 min.
9. Vortex MagMax beads to create uniform suspension.
10. Deliver 10 μL beads to each well.
11. Place plate on orbital shaker, and shake vigorously for 3 min.
12. Move plate to magnetic stand to pellet out beads for 2 min.
13. Aspirate off completely supernatants from pellets, and discard aspirates.
14. Deliver 150 μL enhancer solution to each well.
15. Place plate on orbital shaker, and shake vigorously for 3 min.
16. Move plate to magnetic stand to pellet out beads for 2 min.
17. Aspirate off completely supernatants from pellets, and discard aspirates.
18. Add 150 μL wash buffer II to each well.
19. Place plate on orbital shaker, and shake vigorously for 1 min.
20. Move plate to magnetic stand to pellet out beads for 2 min.
21. Aspirate off completely supernatants from pellets, and discard aspirates.
22. Repeat steps 18–21.
23. Deliver 50 μL TURBO DNase solution to each well.
24. Place plate on orbital shaker, and shake vigorously for 10 min.
25. Add 100 μL lysis/binding buffer to each well.
26. Shake vigorously for 3 min.
27. Move plate to magnetic stand to pellet out beads for 2 min.
28. Aspirate off completely supernatants from pellets, and discard aspirates.
29. Add 150 μL wash buffer II to each well.
30. Place plate on orbital shaker, and shake vigorously for 1 min.
31. Repeat steps 28–30.
32. Aspirate off completely supernatants from pellets, and discard aspirates.
33. Return plate to shaker and shake vigorously for 2 min to dry beads.

34. Deliver 50 μL ethanol to each well, and shake for 3 min.
35. Move plate to magnetic stand to pellet out beads for 2 min.
36. Recover the supernatant as the source of purified RNA.

Key Reference

Ambion Instruction Manual, Version 0506, MagMAX™-96 for Microarrays Kit (Cat. #1839).

Protocol 2.3: Methods for RNA Purification

How It Works

mRNA contains a poly(dA) tail and can be fished out of a cell extract using a solid phase with tethered poly(dT). Classically, this was done with cellulose-dT, in which the cell extract containing RNA was passed over a column containing the cellulose-dT. The RNA was purified away from proteins and other non-poly(A)-tailed RNA and DNA using elution chromatography. Another approach is to use spin columns containing poly(dT) immobilized onto membranes or particle-impregnated membranes. These methods work well for purification of moderate amounts of mRNA from a few samples but are not readily scaleable. For this reason, other solid support formats are more applicable when multiple samples require processing. One approach is to employ a 96-well filter block comprising oligo(dT) membranes. Paramagnetic beads containing poly(dT) also works well and can be scaled up into 96-well or 384-well microplate formats (Figure 2.10) This latter approach is

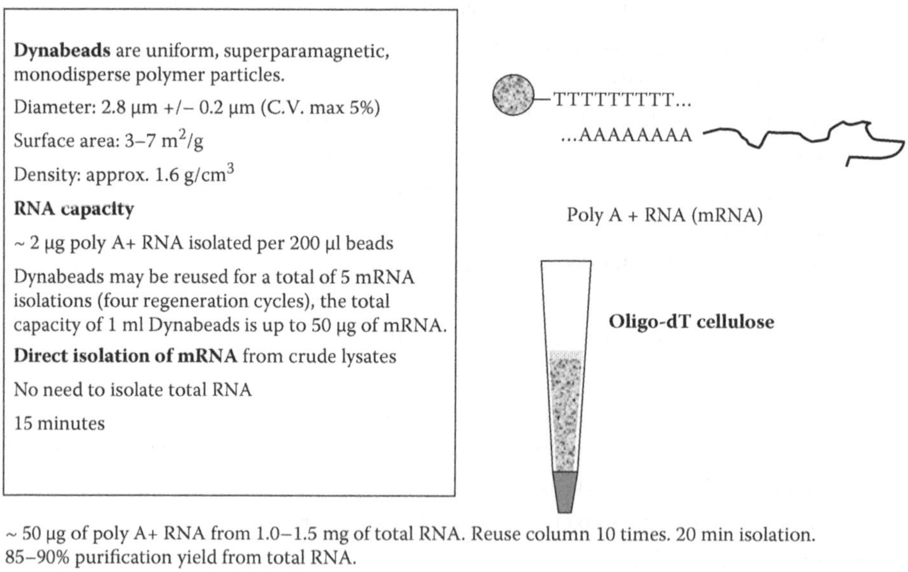

Figure 2.10
Oligo-dT purification of RNA; Dynal's Dynabeads® Oligo (dt)25. (From Molecular Research Center, Inc.)

described in the following text for isolation of mRNA from blood using a manual or an automated approach.

Required Materials

Dynabeads Oligo $(dT)_{25}$ (Dynal)
Magnetic bead capture plate:
Automated method, e.g., MPC-9600 (Dynal, Invitrogen, Carlsbad, CA)
Manual method, e.g., MPC-96B (Dynal, Invitrogen)
Robotic liquid handler, e.g., Biomek 2000 Laboratory Automated Workstation (Beckman Coulter, Fullerton, CA)
Tube rotator
Tube heater block

Reagent Preparation

Wash buffer: 10 mM Tris-HCl, 1 mM EDTA, pH 8.0 containing 0.15 M LiCl
LiDS wash buffer: 10 mM Tris-HCl, 1 mM EDTA, pH 8.0 containing 0.15 M LiCl, 0.1% LiDS
Lysis buffer: 100 mM Tris-HCl, 10 mM EDTA, pH 8.0 containing 500 mM LiCl, 1% LiDS, 5 mM dithiothrietiol
Elution buffer: 2 mM EDTA, pH 8.0
Regeneration solution: 0.1 M NaOH

Protocol 2.3.1: Manual Isolation of mRNA from Tissue

Step-by-Step Protocol

1. Resuspend the beads prior to use.
2. Pipette 250 μL bead suspension into an Eppendorf tube loaded into an MPC holder.
3. Let stand for 30 s.
4. Pull off the cleared supernatant from the tube with a pipette and discard. Note: Avoid disturbing bead pellet.
5. Remove tube from MPC holder.
6. Resuspend beads in 200 μL lysis buffer.
7. Place tube back in MPC holder.
8. Homogenize 25 mg tissue in an Eppendorf tube with 1 mL lysis buffer.
9. Centrifuge lysate for 30 s.
10. Recover supernatant into new tube.
11. Reduce viscosity—take up lysate supernatant into 2 mL syringe equipped with 21 gauge needle and mix three times up and down into tube to shear DNA.
12. Return to step 7—pull off lysis buffer from beads.
13. Remove bead tube from MPC holder.

14. Add lysate from step 11 to bead tube.
15. Place bead-lysate tube on rotator for 5 min.
16. Return bead-lysate tube to MPC holder.
17. Let stand for 2 min.
18. Pull off supernatant with pipette.
19. Remove bead-lysate tube from MPC holder.
20. Add 1 mL LiDS wash buffer and mix well. Place back on MPC holder, and pull off supernatant and discard.
21. Repeat steps 19–20.
22. Remove bead-lysate tube from MPC holder.
23. Add 0.5 mL wash buffer and mix well. Place back on MPC holder, and pull off supernatant and discard.
24. Remove bead-lysate tube from MPC holder.
25. Add 20 μL elution buffer and mix well.
26. Place bead-lysate tube in heater block for 2 min at 65°C.
27. Return bead-lysate tube to MPC holder.
28. Remove supernatant (20 μL) and SAVE as source of mRNA.

Key Reference

Dynabead Oligo $(dT)_{25}$-Package Insert (Dynal), Chapter 2: mRNA Isolation Using Dynabeads Oligo $(dT)_{25}$.

Protocol 2.3.2: Automated Isolation of mRNA from Whole Blood

Step-by-Step Protocol

1. Set up Biomek 2000 worksurface (deck) equipped with:
   a. Multichannel pipetting tools, MP-20 and MP-200
   b. P200 tip rack (2) and P20 tip rack (2)
   c. MPC-9600
   d. Gripper tool
   e. Tube holder, 1.2 mL Eppendorf tubes
   f. Tube holder, 200 μL Eppendorf tubes (PCR tubes)
   g. Waste reservoir
   h. 96-Well plate containing blood samples
2. Manually pipette Dynabeads into 1.2 mL tube on deck.
3. Manually pipette lysis, wash, and elution buffers into 1.2 mL tubes on deck.
4. Transfer 200 μL beads using MP-200 into 1.2 mL tubes.
5. Using a gripper tool, move tubes containing beads to MPC-9600.
6. Let stand in MPC holder for 30 s.

7. Remove supernatants using the MP-200 tool.
8. Using the gripper tool, move tubes off the MPC holder.
9. Pipette 200 μL lysis buffer into tubes containing beads.
10. Transfer 10 μL blood sample using MP-20 into PCR tubes.
11. Transfer 200 μL beads in lysis buffer using MP-200 into PCR tubes containing blood sample.
12. Pause program for 5 min.
13. Move bead tubes back onto MPC holder using gripper tool.
14. Let stand in MPC holder for 30 s.
15. Aspirate off supernatants, and discard using MP-200.
16. Using the gripper tool, move tubes off the MPC holder.
17. Pipette 200 μL LiDS wash buffer into tubes.
18. Return tubes to MPC holder.
19. Aspirate off supernatants, and discard using MP-200.
20. Repeat steps 15–19 twice more.
21. Repeat steps 15–19 with wash buffer.
22. Pipette 40 μL elution buffer into tubes, and resuspend bead pellet using the MP-200.
23. Let stand in MPC holder for 30 s.
24. Remove supernatant using MP-200 and transfer to fresh 200 μL tube.
25. SAVE supernatant as source of mRNA.

Key Reference

Merel, P. et al. (1996). Completely automated DNA extraction from whole blood. *Clinical Chemistry,* 42: 1286–1286.

Protocol 2.4: Electrophoresis of Nucleic Acids

Protocol 2.4.1: RNA Denaturing Agarose Gel Electrophoresis

How It Works

Following the extraction and isolation of total RNA, its quality needs to be assessed. Agarose (1–2%) gel electrophoresis is a quick and easy method of determining the quality of your sample (Figure 2.11). However, RNA has very significant secondary structural effects such as hairpin loops and backfolding, and therefore does not migrate in a sieving gel according to size. To overcome this problem, the best approach is to run RNA under denaturing conditions. This protocol prepares a ~1.2% agarose gel containing 1.9% formaldehyde (0.67 M).

Required Materials

Agarose
Formamide (deionized)

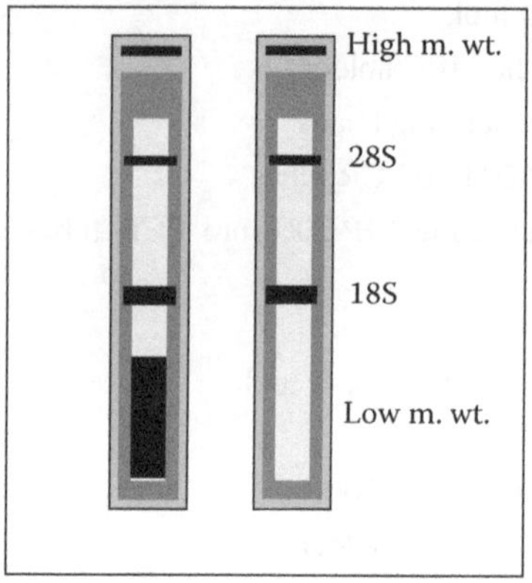

- RNA Purity
  - Agarose Gel Electrophoresis
    - >500 ng total RNA
    - 1% agarose
    - Degradation
      - Low intensity 28S
      - Appearance of low m. wt. smear
  - RNA Spectral Scan
    - A260/A280 = 1.9–2.0
  - Agilent 2100

Figure 2.11
RNA purity checks.

Formaldehyde (37% aqueous; 13.3 M at specific gravity ~1.08)
MOPS
Sodium acetate
EDTA, disodium (Na_2EDTA)
NaOH (1 M)
DEPC water
Glycerol
SYBR Green
Loading dye (bromophenol blue, BPB)
Erlenmeyer flask (rinsed in DEPC water)
Horizontal (submarine) gel electrophoresis apparatus
Power supply
Fume hood
Balance
Microwave

Reagent Preparation

MOPS (10X) gel buffer: To prepare 1 L, dissolve 41.86 g MOPS, 6.81 g sodium acetate, and 3.72 g Na_2EDTA in 900 mL DEPC-treated water. Adjust to pH 7.0 with 1 M NaOH, then bring the volume to 1 L with water.

Gel running buffer: To prepare 1 L, dilute 100 mL MOPS (10X) into 882 mL DEPC water. Add 18 mL formaldehyde and mix well.

Loading buffer (5X): To prepare 1 mL, weigh out ~2–3 mg bromophenol blue (solid) and transfer into a 1.5 mL Eppendorf tube. Deliver 400 μL running buffer (10X), 200 μL glycerol, and 290 μL DEPC water. Mix well, then add 30 μL formamide; 8 μL 500 mM EDTA, pH 8.0; and 72 μL of formaldehyde. Mix and then store at 4°C.

Step-by-Step Protocol

1. Weigh out 1.2 g of SeaKem (or similar quality) agarose, and place in a 250 mL Erlenmeyer flask.
2. Add 85 mL DEPC-treated water.

3. Cover flask with loosely fitting cap (or paper towel) to prevent evaporation.
4. Heat the flask in a microwave for 45 s at maximum power level (e.g., setting = 10).
5. Swirl to dissolve any clumps, then heat for 1 min at low power level (e.g., setting = 3).
6. Place heated flask in water bath at 60–65°C for at least 5 min to cool solution prior to additions and pouring. Check solution temperature with thermometer.
7. Set up remaining steps in a fume hood.
8. Add 10 mL MOPS (10X) gel buffer.
9. Add 5.0 mL formaldehyde to flask, and swirl to mix into solution.
10. Set up the gel box with loading comb.
11. Pour agarose gel into box, and cool to solidify gel.
12. Remove comb, and prepare samples for loading.
13. Heat 1–2 μg RNA sample in Eppendorf microtube for 10 min at 75°C.
14. Transfer tube onto ice.
15. Add denaturant and loading dye to sample tube.
16. Load samples and molecular weight standards into gel wells.
17. Submerge gel with running buffer.
18. Turn power on, and electrophoresis at 100 volts for 1–2 h or until running dye migrates to bottom portion of gel.
19. Remove gel from gel box, and place in staining tray.
20. Add SYBR Green solution, and stain for several minutes to develop bands.

Key References

Qiagen Corporation (2001). Bench Guide. Chapter 5, RNA: A guide to analytical gels.

University of Hawaii, Advances in Bioscience Education: http://abe.leeward.hawaii.edu/Protocols.htm.

Protocol 2.4.2: Slab Gel Electrophoresis of Extracted and Amplified DNA Products

How It Works

Electrophoretic analysis of extracted DNA, amplified DNA (amplicons), or cDNA requires less demanding conditions and setup. Approximately 1–2% agarose cast in a submarine gel using nondenaturing gel buffer is used. DNA in the range of ~50–1000 bp is separated.

Required Materials

Agarose
Ethidium bromide or Syber Green
DNA ladder
Loading dye (bromophenol blue, BPB)
Gel buffer (Tris-borate-EDTA, TBE)
Horizontal (submarine) gel electrophoresis apparatus

Power supply
Balance
Microwave

Step-by-Step Protocol

1. Weigh out 0.32–0.33 g SeaKem LE or similar-quality agarose.
2. Add 25 mL 1X TBE.
3. Heat in microwave oven for 45 s, power level 10 (max).
4. Swirl to dissolve any clumps, then heat for 1 min + 10 s at power level 3.
5. Place in 48°C water bath for at least 5 min to cool.
6. Add ~1 µL of ethidium bromide solution (10 mg/mL water).
7. Pour agarose in submarine gel holder with well comb and allow to cool (15 min) to cast gel; remove air bubbles with Kimwipes and level gel as required.
8. Load 10 µL 100 bp DNA ladder containing tracking dyes into outer wells. (Note: Do not use outermost wells.)
9. Load 5–10 µL PCR reaction mixture into wells.
10. Fill remaining wells with 10 µL TBE buffer.
11. Start voltage @ 270 V to move sample into gel, allowing dyes to separate (~5 min).
12. Reduce voltage to ~85 V until first dye migration reaches near the bottom third of the gel (~45 min)

Protocol 2.5: Labeling of Nucleic Acid Targets

Protocol 2.5.1: Aminoallyl dUTP Incorporation into cDNA

How It Works

This is a postsynthesis, nonenzymatic dye labeling methodology based on the use of the amine-reactive *N*-hydroxysuccinimidyl esters of Cy3 and Cy5. Amine-reactive groups are incorporated into cDNA strands during reverse transcription of RNA, using 5-(3-Aminoallyl)-2′-deoxyuridine 5′-triphosphate, commonly referred to as aa-dUTP. The process reduces the likelihood of excessive incorporation of dye, which often leads to fluorescence quenching, thereby reducing the overall signal strength. Thus, the aminoallyl-mediated dye-labeled nucleic acids produce more intense signals on microarrays (Figure 2.12).

Required Materials

5-(3-Aminoallyl)-2′-deoxyuridine 5′-triphosphate sodium salt
N-hydroxysuccinimidyl Cy3
N-hydroxysuccinimidyl Cy5
dNTPs
Reverse transcriptase, e.g.
  SuperScript III (Cat. No. 18080-093, Invitrogen) supplied at 200 U/µL includes:
  1st Strand Buffer (5X): 250 mM Tris-HCl, pH 8.3, 375 mM KCl, 15 mM $MgCl_2$
  DTT (0.1 M)

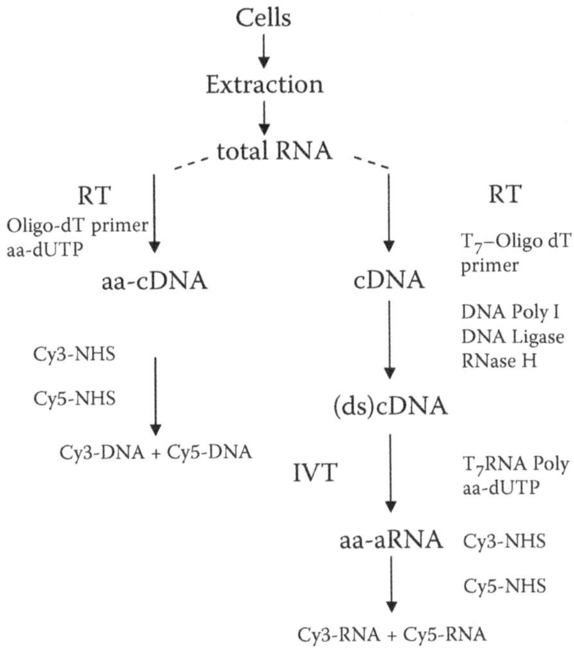

Figure 2.12
Labeling RNA. (Adapted from Ross et al., 2000.)

HEPES buffer, pH 7.0
Oligo-d$(T)_{20}$, 50 μM
Random primer mixture, 9-mers
Poly (A)+ RNA, 2 μg per reaction
Thermocycler
200 mM NaOH–20 mM EDTA stock solution
Spin filter (size-exclusion type)

Reagent Preparation

dNTP/aa-dUTP stock solution (50X): Mix 50 μL each of 100 mM dATP, dCTP, dGTP + 25 μL dTTP (100 mM), and 50 μL aa-dUTP (50 mM).

cDNA reaction mix (freshly prepared in 100–200 μL Eppendorf tube). The following is enough to conduct two synthesis reactions. Typically, Cy3-labeled reference (control) and Cy5-labeled test (experimental) samples are prepared. For two reactions, mix:

1st Strand Buffer	5X	13.8 μL
aa-dUTP	50X	1.4 μL
DTT	0.1 M	6.9 μL
SuperScript	200 U/μL	4.6 μL
Water		3.5 μL
Total Volume		30.2 μL

Step-by-Step Protocol (Per Single Reaction)

Incorporation of aa-dUTP into cDNA

1. In an Eppendorf tube (1.5 mL capacity), add 2 μg RNA sample along with 1 μL oligo(dT) primer + 1 μL random primer. Adjust the final volume to about 15 μL with water.
2. Close cap, and incubate 10 min at 70°C in a thermocycler.
3. Remove the tube, and place immediately on ice for 5–10 min.
4. Open tube, and add 15 μL cDNA reaction mix. Mix well by vortex.
5. Close cap, and incubate tube for 2 h at 42°C in a thermocycler.
6. Incubate 5 min at 95°C.
7. Open tube, and add 30 μL of NaOH–EDTA denaturing solution. Mix well by vortex.
8. Close cap, and incubate tube for 15 min at 67°C in a thermocycler.
9. Remove tube, and allow to cool to room temperature.
10. Open tube, and add 60 μL 1 M HEPES, pH 7.0 to neutralize.
11. Adjust final volume to 500 μL with water (~380 μL). Mix well by vortex.
12. Transfer the contents of the tube to a size-exclusion spin filter.
13. Concentrate volume to approximately 10–20 μL by spinning filter for 10 min.
14. Adjust filtrate volume to 500 μL, and repeat step 12. Try to concentrate the filtrate to approximately 10 μL, but avoid drying the filter.
15. Remove the concentrated filtrate (aa-cDNA), and store at −20°C until needed for labeling.

Preparation of Dye-Labeled cDNA

16. To 10 μL of the aa-cDNA prepared in step 14 (above), add 10 μL of 1 M sodium bicarbonate, pH 9.
17. Open single-use NHS-dye packet (e.g., NHS-Cy3 or NHS-Cy5), and add the aa-cDNA bicarbonate solution just prepared in step 1 (Part II).
18. Mix well by vortex.
19. Cover tube with aluminum foil, and incubate for 1 h at ambient temperature.
20. Open tube and add DNA binding buffer.
21. Transfer contents of the tube to a DNA binding spin filter (e.g., used for PCR amplicon purification).
22. Wait 10 min, and then spin column to remove effluent. DNA should be now bound to column matrix.
23. Apply rinse buffer to column, and spin to wash out uncoupled dye and other reagents.
24. Repeat step 8, twice more.
25. Add elution buffer, incubate for 10 min, and then spin column to recover dye-labeled aa-cDNA in effluent tube.
26. Concentrate the eluate by speed-vac.
27. Store at 4°C until needed for hybridization to microarray.

Key References

DeRisi, J. (2001). Amino-allyl dye coupling protocol. Updated version from Brown Laboratory at Stanford University; Meyers et al. (2006). http://cmgm.stanford.edu/pbrown/protocols/RTaminoAllylCoupling.html.

Hegde, P. et al. (2000). A concise guide to cDNA microarray analysis. *BioTechniques,* 29: 548–562.

Ross, D. T. et al. (2000) Systematic variation in gene expression patterns in human cancer cell lines. *Nature Genetics*, 24: 227–235.

Protocol 2.5.2: Dye Incorporation into mRNA

How It Works

This method incorporates what is commonly referred to as the Eberwine protocol for labeling mRNA (Figure 2.13). The method avoids amplification pitfalls associated with PCR involving Taq polymerase. In particular, it has been recognized that Taq is relatively inefficient at the amplification of long sequences of target DNA. Because of this, a portion of the amplified population of cDNA is of smaller size and is not fully representative of the original target population distribution. The Eberwine protocol is thought to produce a more representative population. This is crucial in comparing two populations of expressed targets.

The original Eberwine protocol is accomplished as follows. A heterogeneous pool of cDNA is prepared by reverse transcription of total RNA (sample target) using a poly(dT) primer tailed with a T7 promoter sequence. Following 2nd strand

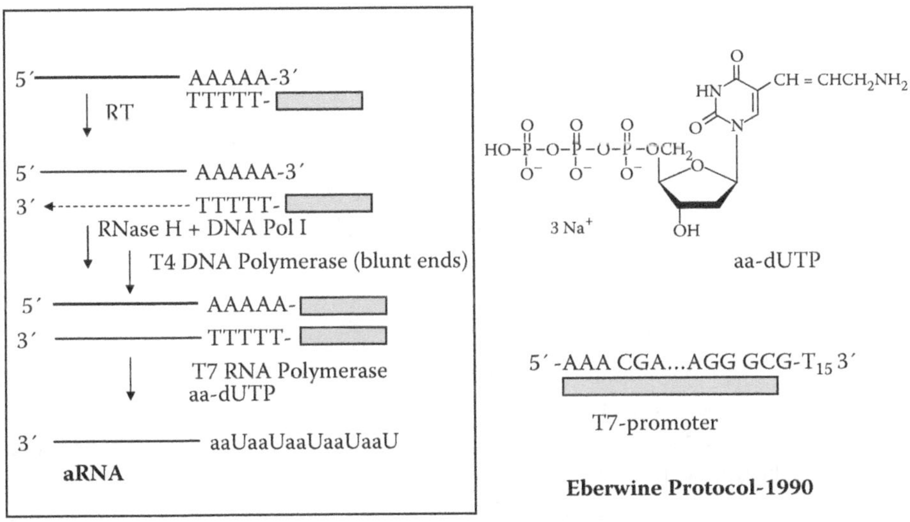

Figure 2.13
Amplification of mRNA for labeling. (Modified from van Gelder et al., 1990.)

synthesis of cDNA by RNaseH-DNA polymerase I, a T7 RNA polymerase is introduced, which catalyzes the synthesis of the corresponding antisense RNA strand from the ds cDNA template. The amplified product, called amplified antisense RNA, or aRNA, with incorporated label ^{32}P-CTP, biotin-CTP (or Cy3/Cy5-CTP) is used as the labeled target source of hybridization to microarrays. Alternatively, aa-UTP can be incorporated and the RNA postlabeled with Cy3/Cy5-NHS.

Required Materials

Total RNA, 5–40 μg
T7-dT15 primer: 5′-AAA CGA CGG CCA GTG AAT TGT AAT ACG ACT CAC TAT AGG CGC- T_{15}-3′
Reverse transcriptase, AMV-RT or MLV RT or SuperScript II RT (200U/μL)
RNase H
DNA polymerase I
T4 DNA polymerase
Thermocycler

Reagent Preparation

aRNA amplification cocktail: 6 mM MgC12; 10 mM NaCl; 2 mM spermidine; 10 mM dithiothreitol; 500 μM (each) ATP, GTP, UTP; 12.5 μM CTP + 30 μCi of [α-^{32}P]CTP; 10 units, RNase block; 80 units T7 RNA polymerase.

Step-by-Step Protocol

1st and 2nd Strand cDNA Synthesis

1. Add to a 0.5 mL Eppendorf tube (nuclease-free) the following template mix per tube:
   a. 2 μL total RNA (10 μg)
   b. 1 μL T7-dT15 primer (100 pmoles)
   c. 9 μL DEPC water
2. Place tube in thermocycler (or heating block; water bath) and incubate for 10 min at 70°C.
3. Centrifuge tube, and place on ice for 5 min.
4. Add the following as a master mix containing per tube:
   a. 4 μL 5X RT buffer (1st strand buffer supplied with enzyme)
   b. 1 μL 10 mM dNTP's
   c. 2 μL 100 mM DTT
5. Preincubate tube for 2 min at 42°C.
6. Add 1 μL SuperScript to tube, mix well, and incubate for 1 h at 42°C.
7. Place tube on ice for 5 min.
8. Prepare a 2nd Strand Master Mix, comprising per each addition to a 1st Strand tube from step 7, the following:
   a. 30 μL 5X 2nd Strand Buffer
   b. 3 μL 10 mM dNTP's
   c. 1 μL DNA ligase

   d. 1 μL RNase H
   e. 4 μL DNA Poly I
   f. 91 μL DEPC water
9. Remove tube (step 7) from ice, centrifuge to collect condensation, and add the 2nd strand mixture.
10. Incubate tube at 16°C for 2 h in a circulating cold water bath.
11. Add 2 μL of T4 Polymerase, and incubate for an additional 5 min at 16°C.
12. Terminate reaction by the addition of 10 μL 0.5 M EDTA.
13. Store ds cDNA at –20°C until ready for use or proceed to cDNA cleanup.

ds cDNA Cleanup

14. Conduct extraction directly from the cDNA reaction tube (step 13, above) by adding an equal volume (162 μL) of the phenol:chloroform: isoamyl alcohol solution.
15. Mix the tube by vortex for 1 min.
16. Centrifuge at 12,000× g for 2 min.
17. Remove the aqueous layer, and transfer to new 1.5 mL Eppendorf tube. Discard organic layer.
18. Add ½ volume (80 μL) of 7.5 M ammonium acetate.
19. Add 2.5 volumes (400 μL) of cold (–20°C) 100% ethanol.
20. Vortex tube, then centrifuge at 12,000× g for 20 min to pellet DNA.
21. Remove supernatant and discard.
22. Add 0.5 mL of cold 80% ethanol, and vortex.
23. Centrifuge at 12,000× g for 5 min.
24. Gently remove the supernatant, avoiding dislodging of the pellet.
25. Repeat steps 9–12 once again.
26. Allow the pellet in the tube to air-dry.
27. Resuspend the pellet in a small volume of RNase-free water or TE. Note: The volume used will depend on the required amounts for the IVT reaction, typically about 1–5 μg ds cDNA.

In Vitro Transcription: Generation and Labeling of aRNA

28. Prepare an NTP Master Mix containing equivalent volumes of ATP, CTP, GTP and UTP (75 mM each), and 10X reaction buffer:
   a. 2.5 μL *each* NTP
   b. 2.5 μL reaction buffer (10X)
29. Prepare template in a 0.1 mL nuclease-free Eppendorf tube, containing:
   a. 5 μL ds cDNA
   b. 2 μL T7 RNA polymerase
   c. 3 μL nuclease-free water
30. Transfer 10 μL of the NTP mix into the template tube, and gently mix.
31. Incubate for 4 h at 37°C in a thermocycler.
32. Add 1 μL DNase I, mix, and incubate for an additional 15 min at 37°C.

33. Place on ice for 5 min, then centrifuge to collect condensate.
34. Store at –20°C.

Key References

Expression Analysis Technical Manual, Chapter 2, pp. 9–11 (P/N 700218 rev. 2, Affymetrix, 1999).

http://www.usc.edu/schools/medicine/research/institutes/igm/content/microarray/protocols.htm.

Ma, C. Q. et al. (2006). *In vitro* transcription amplification and labeling methods contribute to the variability of gene expression profiling with DNA microarrays. *Journal of Molecular Diagnostics,* 8(2): 183–192.

Van Gelder, R. N. et al. (1990). Amplified RNA synthesized from limited quantities of heterogenous cDNA. *Proceedings of National Academy of Sciences USA,* 87: 1663–1667.

Protocol 2.6: Storage of Nucleic Acids

Protocol 2.6.1: RNA

Long-term storage of purified RNA at –70°C to –80°C in water for 1 year is recommended. Avoid freeze–thaw cycling. For best results, aliquot and use once. You may store RNA at –20°C for several months. Freeze-drying of RNA in the presence of 10% trehalose has also been shown to preserve RNA at 4°C (Jones et al., 2007).

Protocol 2.6.2: DNA

Genomic DNA that has been prepared DNAse free may be stored for at least 1 year at –70°C to –80°C in TE buffer. The addition of TE buffer is recommended to prevent further nuclease activity. DNA stored in water for an extended time may degrade from acid hydrolysis, and therefore, buffering is recommended. Isolated DNA stored in TE may be stored at 4°C for several days and 20°C for several months. DNA that has been dried down or adsorbed onto FTA paper (Whatman) is very stable and can be stored for years at room temperature (Rogers and Burgoyne, 1997).

Protocol 2.6.3: Primers and Probe Oligonucleotides

Oligonucleotides are best stored lyophilized until needed. However, certain labeled oligonucleotides are best stored in TE buffer in order to maintain solubility and avoid aggregation. Stock solutions of primers may be stored at –70°C to –80°C. Aliquoting is highly recommended. Working solutions of oligonucleotides can be preserved at –20°C for 1 year or at 4°C for several months. Avoid handling any stocks without wearing gloves in order to reduce nuclease contamination or introduction of PCR target cross-contamination.

Protocol 2.6.4: Activated Oligonucleotides

Synthetic oligonucleotides can be activated with moisture-sensitive, labile groups for coupling to solid supports or other biomolecules. For example, this applies to oligonucleotides that have been synthesized with terminal reactive groups such as thiols (-SH) or activated esters (-NHS). These are to be stored lyophilized and sealed under nitrogen at –70°C to –80°C until needed. To use, bring to room temperature and resuspend in the appropriate coupling buffer. Use immediately or place in a stabilizing media just prior to use, such as an inert, dry organic solvent (DMF; DMSO), depending on the constraints of the intended coupling reaction.

Key References

Jones, K. L. et al. (2007). Long-term storage of DNA-free RNA for use in vaccine studies. *BioTechniques,* 43(5): 675–681.

Rogers, C.D. and Burgoyne, L.A. (1997). Bacterial typing: Storing and processing of stabilized reference bacteria for polymerase chain reaction without preparing DNA—An example of an automatable procedure. *Analytical Biochemistry,* 247: 223–227.

Chapter 3

Solid-Phase Substrates for Nucleic Acid Microarrays

Robert S. Matson

Contents

Introduction

Substrate Selection

The most widely adopted substrate for the preparation of nucleic acid-based microarrays remains the glass microscope slide to which various solid phases have been applied. Although a number of different solid phases have been developed for attaching nucleic acid probes, only a few have been commercialized. Glass quality is important for use with confocal scanners. Specific tolerances in flatness and the magnitude of autofluorescence must be taken into account. Glass cleanliness is also critical for uniform coverage of the solid phase, as well as for reduction of fluorescing particles.

Immobilization by Adsorption

A nucleic acid whose backbone contains phosphate linkages can be considered a polyanion or negatively charged macromolecule. For that reason, DNA can be easily adsorbed to a cationic surface by forming salt (ionic) bridges. Thus, DNA can be printed down onto the positively charged poly-L-lysine-coated slides with relative ease (Figure 3.1).

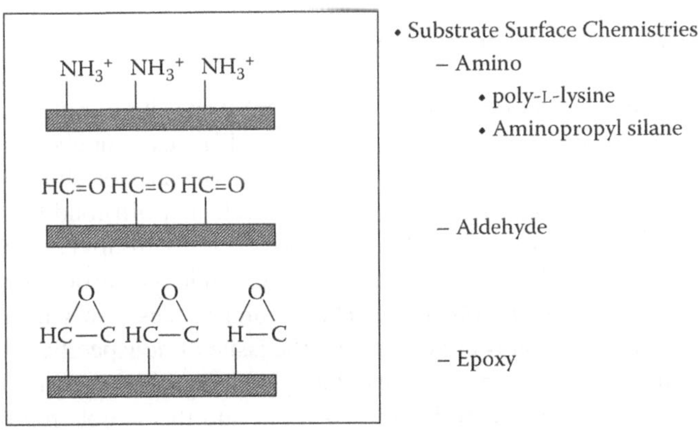

Figure 3.1
Substrate surface chemistries.

Covalent Attachment

If, in fact, DNA is easily immobilized by passive adsorption, why resort to using a covalent chemistry for attachment? There are really two rationales for doing so. First, the efficiency for adsorptive-driven electrostatic binding of DNA to a charged surface is largely related to size and structure. That is, a larger DNA fragment contains more phosphate groups than a smaller fragment, and that increases the likelihood of greater binding. On the other hand, we know that folding, hairpin loops, etc., can sterically hinder binding. Therefore, secondary structure also comes into play here and must be taken into consideration when designing probes and performing hybridization. However, it is generally understood that the immobilization of higher molecular weight, for example, cDNA, by adsorption is readily accomplished, whereas adsorption of low molecular weight, for example, oligonucleotide probes, is more problematic. Thus, short, single-stranded probes are best immobilized by covalent means. This brings into focus the second reason for taking the route of covalent attachment: hybridization efficiency and selectivity. Shorter capture probes are thermodynamically designed to be most useful for detection of closely related nucleic acid targets, such as in the case of mutation analysis or single-nucleotide polymorphism (SNP) screening. Under these conditions, the participation of the entire sequence is important in order to quantitatively differentiate between the wild-type (normal) and mutant or polymorphic species. The approach most often taken is to tether probes at their 5′ terminus, which permits efficient hybridization of the entire sequence. This is accomplished by chemically modifying the 5′ end with a primary amine group. Because primary amines are much more reactive than secondary amines found in the nucleotide bases, the probe is attached and oriented by its 5′ terminus. Those requiring further details on covalent attachment of oligonucleotides to solid-phase supports should review the treatise by Serge Beaucage (2001).

Quenching, Capping, and Blocking

Once the capture probe is immobilized to the substrate, it is important to perform two additional steps prior to using the microarray. If a covalent chemistry was used for immobilization, any residual reactive groups on the surface should be removed. This is commonly called *quenching* the surface. Under certain conditions, this is also referred to as *capping*. For example, residual epoxide (EP) groups can be reacted with an amine compound such as ethanolamine, whereas aldehyde groups can be reduced to alcohols using sodium borohydride. The second process we will call *blocking*. Once residual reactive groups are destroyed, the issue of nonspecific adsorption will need to be addressed. What you choose to block with depends on several factors such as the treated surface, the hybridization cocktail, and the sample matrix. Common blocking agents include detergents such as Tween 20, salmon sperm DNA, tRNA, or proteins such as bovine serum albumin (BSA).

Hybridization

Once you have properly prepared the microarray, you are ready to apply the sample. The sample is generally derived from a biological matrix such as a cell extract, tissue, plasma, etc., and contains both the target and nontarget genomic nucleic acid species. The good news is that we can generally extract, purify, and amplify (e.g., polymerase chain reaction [PCR]) the target from the sample in advance of hybridization. There are also high-throughput automated methods for doing so for multiple samples. However, this is not always possible, especially with clinical specimens. It may be necessary to include additives to the hybridization buffer to reduce nonspecific hybridization. Hybridization is a thermodynamic process with base-pairing rules that permit proper duplex formation between the solid-phase tethered probe and the single-stranded target. Use the nearest-neighbor approximation model to design probes and develop a hybridization strategy. Common hybridization buffers include SSC (saline, sodium citrate), SSPE (saline, sodium phosphate, EDTA) with additives such as formamide that reduce differences in melt temperature (T_m) due to the occurrence of strong base–pair interactions. Other additives would include accelerating agents such as dextran sulfate, and the inclusion of tRNA to reduce interferences from genomic DNA.

List of Protocols

Protocol Number	Name
3.1	Evaluation of Microarray Substrates for Nucleic Acid Analysis
3.2	Noncovalent Adsorption of DNA to Amino-Silane Supports
3.3	Covalent Attachment
3.4	Quenching and Blocking of Substrates
3.5	Biotinylated cDNA Target Hybridization to cDNA Slide Microarrays

Protocol 3.1: Evaluation of Microarray Substrates for Nucleic Acid Analysis

How It Works

In this protocol, we examine the performance of a microarray substrate for nucleic acid analysis. More importantly, our purpose is to determine which substrate works best for the intended application. How is this accomplished? To do so requires the user to establish boundary conditions on what is an acceptable outcome. For example, for a genotype analysis, the differentiation of closely related genes may be important and requires a 10:1 or greater signal differential between the wild type and a mutation or SNP. For gene expression analysis, reproducibility over a large gene population might be more important. In any case, performance criteria are established and then measured.

Required Materials

cDNA slide array
Petri dish
Temperature-controlled humidity chamber, for example, shaker water bath
Pipettor
Streptavidin conjugate reporter
Scanner or charge-coupled device (CCD) camera system

Reagent Preparation

Denaturant solution (denaturant) = 0.5 M NaOH, 0.15 M NaCl.
Neutralization solution = 2.4× SSC, 0.016% SDS, 0.28 M Tris, 0.028 M NaCl at pH 7.5, prepared by diluting 1.87 mL of 1.5 M Tris, 0.15 M NaCl, pH 7.5 into 8 mL of 3× SSC, 0.02% SDS.
Hybridization solution = cDNA target mixed into the neutralization solution. The final hybridization solution contains oligonucleotide target, 2× SSC, 0.01% SDS, 0.24 M Tris, 0.08 M NaCl, ~pH 8.
2× buffer = 2× SSC, 0.01% SDS.

Step-by-Step Protocol

1. Incubate the microarray substrate, for example, glass slide, containing cDNA probes for 15 min; puddle and then spread out uniformly with 150 μL of denaturant solution to produce single-stranded probes (see Note 1 in the following troubleshooting section).
2. At the end of the 15 min period, rinse the substrate in neutralization solution to neutralize and remove any residual denaturant.
3. Just prior to substrate denaturation (step 1), mix 15 μL biotinylated cDNA with 22.5 μL denaturant in an Eppendorf tube.
4. Incubate for 15 min at room temperature.
5. Next, add 112.5 μL neutralization solution to the tube, and mix well. This becomes the target hybridization (HYB) solution for step 6.
6. Add the HYB solution to the microarray slide (see Note 2 in the following troubleshooting section).
7. Cover the slide and incubate in a humidified chamber overnight at 60°C.
8. Rinse slide in 2× buffer at 60°C.

9. Rinse slide in 2× buffer at room temperature.
10. Add a suitable streptavidin–dye conjugate (e.g., streptavidin-Cy5) or a streptavidin–enzyme conjugate (e.g., streptavidin–alkaline phosphatase).
11. Develop signal, and read on a suitable laser confocal scanner or CCD camera system.

Key References

Beaucage, S. (2001). Strategies in the preparation of DNA oligonucleotide arrays for diagnostic applications. *Current Medicinal Chemistry,* 8: 1213–1244.
Matson, R. (unpublished).

Troubleshooting Guide

Note 1: Slide Denaturation

This step is timed to synchronize with the end of target denaturation (step 3). The surface of a standard microscope slide may be completely wetted out uniformly with this volume.

Note 2: Hybridization

During hybridization, 2× SSC 0.01% SDS buffer is maintained at 60°C inside 15 mL centrifuge tubes in preparation for the stringency rinse. The development process involves a 10 min stringency rinse in 2× SSC 0.01% SDS inside the 60°C shaker. Slides are placed inside a 15 mL centrifuge tube filled to 10 mL with this stringency rinse buffer. A final rinse in this buffer at room temperature completes the process.

Protocol 3.2: Noncovalent Adsorption of DNA to Amino-Silane Supports

How It Works

The original method of creating arrays of nucleic acids on glass slides was based on using poly-L-lysine-coated slides. Adsorption is thought to involve both electrostatic and hydrophobic interaction between the nucleic acid and the polylysine surface. Although such slides are still in use, they have been largely replaced by slides that have been silanized with aminopropyl silane (APS). This surface is considered to be more durable and scalable for manufacture. Generally, the adsorption process is most efficient with higher-molecular-weight nucleic acid polymers, such as PCR products and cDNAs, rather than short-chain oligonucleotides.

Required Materials

PCR amplicons or cDNA
APS slides
Print buffer: 50 mM sodium carbonate–bicarbonate, pH 9.8
DDI water = distilled, deionized (18Ω) water

Reagent Preparation

Print ink: Dissolve amino-oligonucleotides to 20 μM in print buffer
Rinse buffer: 0.2% SDS in print buffer

Step-by-Step Protocol

1. Allow the activated substrate slide to reach room temperature prior to use (see Note 1 in the following section on troubleshooting).
2. Transfer a small volume cDNA stocks into separate wells of a suitable microplate that will be used as the "source plate" for arraying (see Note 2).
3. Remove the source plate from storage, and allow it to reach room temperature prior to use.
4. Centrifuge the source plate at low rpm (<10,000× g) for 5 min to force trapped air out of the wells.
5. Place the source plate on the arrayer deck according to the manufacturer's directions.
6. Open the slides, and carefully place these into position on the deck for arraying (see Note 3).
7. Proceed with printing down the cDNA onto the APS slide in the desired array format.
8. Upon completion of printing, the slides are to be stored under humidity for at least 2 h to allow efficient coupling to occur.
9. Snap-dry the slide at 100°C for 1 min.
10. Rinse the slide in 0.1% SDS.
11. Capping and blocking residual amines (see Note 4)
    a. Optional step: Quench (cap) residual amine residues with a carboxylic acid by acylation using succinic anhydride prepared in 1-methyl-2-pyrrolidinone, boric acid buffer (see Note 5) or
    b. Optional step: Block in 0.1% BSA, 0.1% SDS in 3× SSC for 2 min. Proceed to step 12.
12. Rinse the slides three times in DDI water.
13. Air-dry the slides in a clean, lint-free environment.
14. Store the slides at room temperature in a clean, closed container. Keep dry.

15. Next, slides are placed for 2 min in boiling DDI water (see Note 6). Remove the slides from the boiling water and dry rapidly.
16. Proceed to hybridization.

Key Reference

Taylor, S. et al. (2003). Impact of surface chemistry and blocking strategies on DNA microarrays. *Nucleic Acid Research*, 31(16)e87:1–19.

Troubleshooting Guide

Note 1: Handling Issues

Avoid handling slides with your bare hands. Use gloves at all times. Do not open slides for use in areas where organic amines or ammonia have been recently released into the atmosphere.

Note 2: Source Plates

Usually, a 384-well plate is used with 10–90 μL of the stocks, depending on the requirements of the arraying device. These can be sealed and stored under refrigeration for later use.

Note 3: Use of HEPA Filter

Most arrayers now offer both HEPA filtration and humidity control. We suggest running the HEPA filter just prior to printing for a few minutes to reduce air particulates, and then turning it off. The humidity controller should then be turned on to equilibrate the atmosphere in the arrayer to between 50% and 70% relative humidity.

Note 4: Capping and Blocking Residual Amines

This step may not be necessary, depending on the particular surface chemistry used in preparation. The user should first determine the extent of nonspecific binding in the absence of capping and blocking. It may also be important to explore other blocking agents. We have noted on occasion that the introduction of certain fractions of BSA can lead to an increase in background.

Note 5: Capping Step

Follow Protocol 3.4.1: Capping of Poly-L-Lysine (PLL) Slides using Succinic Anhydride.

Note 6: Denaturation

This process denatures the attached oligonucleotides, thereby rendering them fully single stranded and ready for hybridization.

Protocol 3.3: Covalent Attachment

Protocol 3.3.1: Covalent Coupling of Amino or Other Modified Oligonucleotides to Solid Supports Containing Epoxides (Oxiranes)

How It Works

Although amino-modified nucleic acids may be tethered to the surface of slides or other solid supports using aldehyde chemistry, the additional treatment by borohydride reduction of the Schiff's base is recommended. An alternative surface chemistry is that of epoxide (EP) coupling (see Figure 3.2). The amine-modified oligonucleotide is preferred, although hydroxyl or phosphoryl oligonucleotides may also be coupled, as well as mercapto (-SH)-terminated nucleic acids. It is important that the modified oligonucleotide be present as a strong nucleophile (unpaired set of bonding electrons). This is generally accomplished by placing the oligonucleotide in a strong base, for example, pH 9–10 buffers such as sodium borate or carbonate. Under these conditions, the oligonucleotide can undergo nucleophilic attack of the EP ring, creating a covalent bond. Unlike the Schiff's base, this is a formal covalent bond and is very stable, especially to acidic and alkaline conditions, which is useful during the hybridization and stringency washes.

Required Materials

Modified oligonucleotides: 5′-amino ($-NH_2$), hydroxyl (-OH), mercapto (-SH), and occasionally, phosphoryl ($-OPO_3$) groups are utilized.

Solid support activated with epoxide groups (EP slides).

Print buffer: 50 mM sodium carbonate–bicarbonate, pH 9.8.

DDI water = distilled, deionized (18 Ω) water.

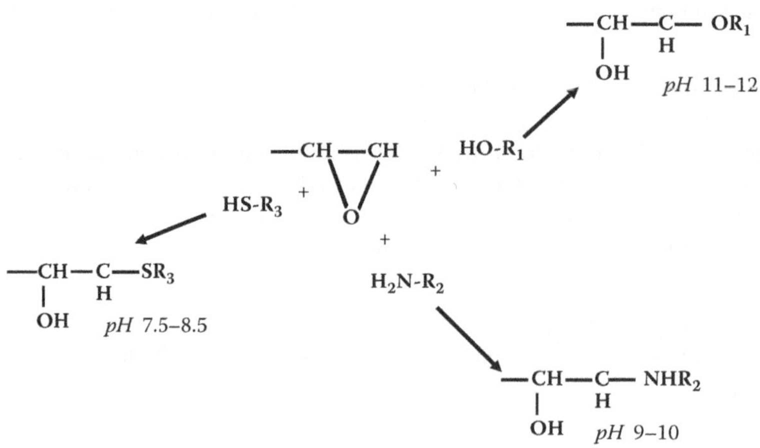

Figure 3.2
Epoxide reaction.

Reagent Preparation

Print ink: Dissolve amino-oligonucleotides to 20 μM in print buffer.
Rinse buffer: 0.2% SDS in print buffer.

Step-by-Step Protocol

1. Allow the activated substrate (EP) slide to reach room temperature prior to use (see Note 1 in the following section on troubleshooting).
2. Transfer a small volume of the 20 μM oligonucleotide stocks into separate wells of a suitable microplate that will be used as the source plate, for arraying (see Note 2).
3. Remove the source plate from storage, and allow it to reach room temperature prior to use.
4. Centrifuge the source plate at low rpm (<10,000× g) for 5 min to force trapped air out of the wells.
5. Place the source plate on the arrayer deck according to the manufacturer's directions.
6. Open the EP slides. Wear gloves at all times to avoid contamination and inactivation.
7. Carefully place these into position on the deck for arraying (see Note 3).
8. Proceed with printing down the oligonucleotides onto the EP slide in the desired array format.
9. Upon completion of printing, the slides are to be stored under humidity for several hours to allow efficient coupling to occur.
10. Air-dry the slides (see Note 4).
11. Rinse the air-dried slides in rinse buffer at least three times with agitation (see Note 5).
12. Rinse the slides three times in DDI water.
13. Quench residual epoxides on EP slides (optional; see Note 6).
14. Air-dry the slides in a clean, lint-free environment.
15. Store the slides at room temperature in a clean, closed container. Keep dry.
16. Next, slides are placed for 2 min in boiling DDI water (see Note 7).
17. Remove the slides from the boiling water and dry rapidly.
18. Proceed to hybridization.

Key Reference

Preininger, C. et al. (2003). Quality control of chip manufacture and chip analysis using expoxy-chips as a model. *Sensors and Actuators* 890, 98–103.

Troubleshooting Guide

Note 1: Handling Slides Avoid handling slides with your bare hands. Use gloves at all times. Do not open slides for use in areas where organic amines or ammonia have been recently released into the atmosphere.

Note 2: Source Plate Usually, a 384-well plate is used with 10–90 μL of the stocks depending on the requirements of the arraying device. These can be sealed and stored under refrigeration for later use.

Note 3: Use of HEPA Filter Most arrayers now offer both HEPA filtration and humidity control. We suggest running the HEPA filter just prior to printing for a few minutes to reduce air particulates, and then turning it off. The humidity controller should then be turned on to equilibrate the atmosphere in the arrayer to between 50% and 70% relative humidity.

Note 4: Drying Arrays Most vendors suggest around 12 h or overnight drying, ensuring a successful good outcome.

Note 5: Washing Slides This can be accomplished by placing the slides in 50 mL conical tubes followed by end-over-end mixing for 5–10 min.

Note 6: Quenching Step The presence of residual surface EPs generally do not present a problem as they are not very reactive under most hybridization conditions. In order for EP-ring opening to occur to an appreciable extent, a strong nucleophile must be employed such as primary amines or thiols under basic pH 9–10 conditions. When in doubt follow Protocol 3.4.3, Quenching of Epoxide Slides, using ethanolamine.

Note 7: Denaturation This process denatures the attached oligonucleotides, thereby rendering them fully single stranded and ready for hybridization.

Protocol 3.3.2: Covalent Coupling of Amino-Oligonucleotides to Solid Supports Containing Aldehydes

How It Works

Oligonucleotides are generally tethered to solid supports by either their 5′ or 3′ ends. The efficiency of coupling is enhanced by introducing nucleophilic groups at the terminus, such as a primary amine group that can react with the activated support. The most common activation is by the introduction of aldehyde groups to the support (Figure 3.3). Primary amines react with aldehyde groups to form a Schiff's base, which represents a semiformal covalent linkage of the nucleic acid to the support. Reduction of the Schiff's base with sodium borohydride results in the more stable amide linkage. Oligonucleotides are prepared in a suitable print buffer that will promote covalent immobilization via the terminal amino group. It is advisable to use a print solution buffered at basic pH ~8–9 so that amino groups serve as good nucleophiles by remaining largely uncharged.

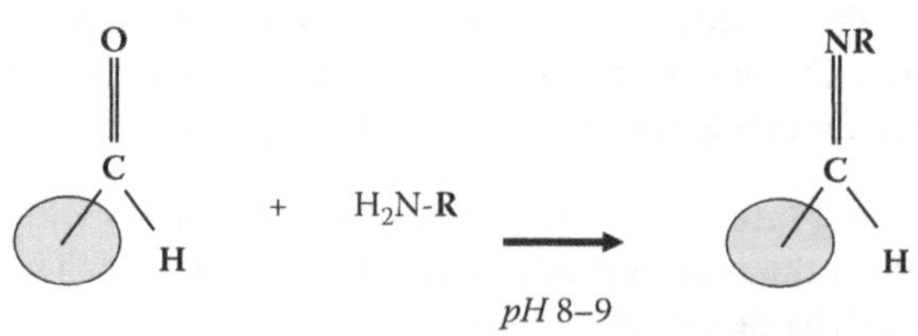

Figure 3.3
Aldehyde coupling reaction, R = nucleic acid/protein/carbohydrate/small molecule.

Required Materials

5′-amino-oligonucleotides.

Solid support activated with aldehyde groups: Pierce, Cat. No. 82054; Super Aldehyde, Cat. No. SMA, ArrayIt.

Print buffer: 3× SSC.

Sodium borohydride ($NaBH_4$) quench solution: Dissolve 1 g $NaBH_4$ in 400 mL PBS, pH 7.2; add 100 mL ethanol and mix well.

DDI water = distilled, deionized (18Ω) water.

Reagent Preparation

Print ink: Dissolve amino-oligonucleotides to 20 μM in print buffer.

Rinse buffer: 0.2% SDS in print buffer.

Step-by-Step Protocol

1. Allow the activated substrate slide to reach room temperature prior to use (see Note 1 in the following section on troubleshooting).
2. Transfer a small volume of the 20 μM oligonucleotide stocks into separate wells of a suitable microplate that will be used as the source plate for arraying (see Note 2).
3. Remove the source plate from storage, and allow it to reach room temperature prior to use.
4. Centrifuge the source plate at low rpm (<10,000× g) for 5 min to force trapped air out of the wells.
5. Place the source plate on the arrayer deck according to the manufacturer's directions.
6. Open the slides, and carefully place these into position on the deck for arraying (see Note 3).
7. Proceed with printing down the oligonucleotides onto the activated slide in the desired array format.
8. Upon completion of printing, the slides are to be stored under humidity for several hours to allow efficient coupling to occur. This is followed by air-drying (see Note 4).
9. Rinse the air-dried slides in rinse buffer at least three times with agitation (Note 5).
10. Rinse the slides three times in DDI water.
11. Air-dry the slides in a clean, lint-free environment.
12. Immerse the slides in $NaBH_4$ solution for 5–10 min in a loosely capped container (Note 6).
13. Dip the slides up and down 20 times into DDI water.
14. Rinse the slides at least three times with DDI water. At this point, the slide is ready for use (step 16) or storage (step 15).
15. Store the slides at room temperature in a clean, closed container. Keep dry.
16. Slides are next placed for 2 min in boiling DDI water (Note 7).
17. Remove the slides from the boiling water, and dry rapidly.
18. Proceed to hybridization.

Key Reference

Iqbal, J. et al. (2008). Fabrication and evaluation of a sequence-specific oligonucleotide mini-array for molecular genotyping. *Indian Journal of Medical Microbiology*, 26: 13–20.

Troubleshooting Guide

Note 1: Handling Slides Avoid handling slides with your bare hands. Use gloves at all times. Do not open slides for use in areas where organic amines or ammonia have been recently released into the atmosphere.

Note 2: Source Plate Usually, a 384-well plate is used with 10–90 μL of the stocks, depending on the requirements of the arraying device. These can be sealed and stored under refrigeration for later use.

Note 3: Use of HEPA Filter Most arrayers now offer both HEPA filtration and humidity control. We suggest running the HEPA filter just prior to printing for a few minutes to reduce air particulates and then turning it off. The humidity controller should then be turned on to equilibrate the atmosphere in the arrayer to between 50% and 70% relative humidity.

Note 4: Drying Arrays Most vendors suggest around 12 h or overnight drying, ensuring a successful good outcome.

Note 5: Washing Slides This can be accomplished by placing the slides in 50 mL conical tubes followed by end-over-end mixing for 5–10 min.

Note 6: Gas Liberation The reaction liberates hydrogen gas, which can build up in a tightly capped container. The reagent is exhausted within 1 h.

Note 7: Denaturation This process denatures the attached oligonucleotides, thereby rendering them fully single stranded and ready for hybridization.

Protocol 3.4: Quenching and Blocking of Substrates

Protocol 3.4.1: Capping of Poly-L-Lysine (PLL) Slides Using Succinic Anhydride

How It Works

Oligonucleotides or cDNA are first printed down onto a glass slide coated with PLL. The adsorbed nucleic acid probes are then UV-cross-linked, followed by baking at 80°C. Residual lysine residues are capped with the addition of succinic anhydride. Subsequently, the probes are denatured at 95°C to render them single stranded, and then dried for storage.

Required Materials

PLL-coated slides
Succinic anhydride (anhydrous)
1-methyl-2-pyrrolidinone (MP)
Sodium borate (1 M, pH 8.0)
Ethanol (95%)
Screw cap conical tubes (50 mL, nominal)
Nitrogen gas
Slide centrifuge

Reagent Preparation

Succinic anhydride (0.16 M) capping solution: Completely dissolve 2 g succinic anhydride into 117 mL MP. Add 5 mL sodium borate solution and mix well. Use immediately in protocol (see Note 1 in the following section on troubleshooting).

Step-by-Step Protocol

1. Place 2 PLL slides back to back in a conical tube. Prepare 3 conical tubes for the processing of 6 PLL slides.
2. Dispense 40 mL capping solution each into the 3 conical tubes.
3. Immediately replace cap, and mix well.
4. Incubate on an end-over-end rotator for 15 min.
5. Prepare 3 additional conical tubes filled with ~45 mL of DDI water (held at 95°C; see Note 2 in the following section on troubleshooting).
6. Remove slides, drain off excess solution, and then plunge them into the near-boiling water for 90 s.
7. Fill 3 additional conical tubes with ethanol.
8. Remove slides from the hot water.
9. Plunge slides into tubes containing ethanol and cap.
10. Mix by inverting tubes several times.
11. Remove ethanol, and refill with fresh ethanol.
12. Repeat steps 9–11 for a total of three times.
13. Dry slides under a nitrogen gas stream, or place them on a slide rack and centrifuge to dry.
14. Place dried slides in a sealed container with a desiccant pouch.
15. Store at room temperature until needed.

Key References

Rubins, K. (2006). Post-processing of oligo arrays on polyLysine coated slides. From The Brown Laboratory at Stanford University, see http://cmgm.stanford.edu/pbrown/protocols/PolyLysinePostProcess.html.

Taylor, S. et al. (2003). Impact of surface chemistry and blocking strategies on DNA microarrays. *Nucleic Acid Research*, 31(16) e87: 1–19.

Troubleshooting Guide

Note 1: Reagent Preparation It is essential that all of the reagents used be of the highest quality. Succinic anhydride should be stored under desiccation or a fresh unopened bottle used. Methyl pyrrolidinone should be colorless; a yellow coloration indicates oxidation, and it should be redistilled or a fresh bottle obtained.

Note 2: Hot Water Incubate DDI water tubes in a boiling water bath until temperature equilibrated, or carefully pour hot water into the tubes from the bath.

Note 3: Centrifuge A slide centrifuge is a preferred means of drying slides because this avoids the occurrence of smears or water marks across the slides.

Protocol 3.4.2: Capping of APS/PLL Slides

How It Works

An alternative method for end-capping solid-phase amines such as aminopropyl silane (APS)-coated or poly-L-lysine (PLL)-coated slides is acylation. In this case, the amine reacts with the *N*-hydroxysuccinimide (NHS) active ester to form an amide bond.

Required Materials

APS-coated slides
PLL-coated slides (PLL)
N-hydroxysuccinimide (NHS; Pierce, Rockford, IL, Cat. No. 24500)
Sodium bicarbonate (0.1 M, pH 8)
Ethanol (95%)
Screw cap conical tubes (50 mL, nominal)
Nitrogen gas
Slide centrifuge

Reagent Preparation

NHS (0.1 M) capping solution: Dissolve 1.4 g NHS in 125 mL sodium bicarbonate buffer. Use immediately (see Note 1 and Note 2)

Step-by-Step Protocol

1. Place 2 APS/PLL slides back to back in a conical tube. Prepare 3 conical tubes for the processing of 6 APS/PLL slides.
2. Dispense 40 mL capping solution each into the 3 conical tubes.
3. Immediately replace cap and mix well.
4. Incubate on an end-over-end rotator for 1 h.
5. Prepare 3 additional conical tubes filled with ~45 mL of DDI water.
6. Remove slides, drain off excess solution, and then transfer into water tubes.
7. Incubate on an end-over-end rotator for 10–15 min.
8. Fill 3 additional conical tubes with ethanol.
9. Remove slides from water tubes.
10. Place slides into tubes containing ethanol and cap.
11. Mix by inverting tubes several times.
12. Remove ethanol, and refill with fresh ethanol.
13. Repeat steps 9–11 for a total of three times.
14. Dry slides under a nitrogen gas stream, or place them on a slide rack and centrifuge to dry.
15. Place dried slides in a sealed container with a desiccant pouch.
16. Store at room temperature until needed.

Key References

Stern, E. et al. (2006). Electropolymerization on microelectrodes: Functionalization technique for selective protein and DNA conjugation. *Analytical Chemistry*, 78: 6340–6346.

Matson, R. (unpublished).

Troubleshooting Guide

Note 1: NHS Reagent Preparation NHS undergoes hydrolysis under basic pH conditions with a half-life of approximately 1 h at pH 8. For that reason, the reagent needs to be used immediately after preparation.

Note 2: Sulfo-NHS A water-soluble derivative of NHS is sulfo-NHS (Pierce, Cat. No. 24510). It can be used in place of NHS if desired. However, it is subject to similar rates of hydrolysis and costs considerably more.

Protocol 3.4.3: Quenching of Epoxide Slides

How It Works

Amino-modified oligonucleotide probes that are arrayed on EP slides are immobilized by a nucleophilic attack of the epoxide resulting in ring opening and the formation of an alkylamine bond. Residual epoxide groups in the presence of ethanolamine are likewise opened and capped by hydroxyl groups (see reaction).

Required Materials

EP slides
Ethanolamine (0.2 M) (see Note 1 in the following section on troubleshooting)
HCl (1 M)
SDS (10% v/v)
DDI water
Screw cap conical tubes (50 mL, nominal)
Nitrogen gas
Slide centrifuge

Reagent Preparation

Ethanolamine (0.2 M) stock: Dilute 12 mL ethanolamine (reagent-grade) to 1 L with DDI water.

Ethanolamine (0.1 M, pH 9) quenching solution: Prepare by titration of 100 mL, 0.2 M ethanolamine with dilute HCl to pH 9. Add 2 mL SDS (10%) just prior to use. Adjust final volume to 200 mL with DDI water. Mix well (Note 2).

Step-by-Step Protocol

1. Place 2 EP slides back to back in a conical tube. Prepare 5 conical tubes for the processing of 10 EP slides.
2. Dispense 40 mL quench solution each into the 5 conical tubes.

3. Replace cap, and mix well.
4. Incubate on an end-over-end rotator for 1 h.
5. Prepare 5 additional conical tubes filled with ~45 mL of DDI water.
6. Remove slides, drain off excess solution, and then transfer into water tubes.
7. Incubate on an end-over-end rotator for 10–15 min.
8. Fill 5 additional conical tubes with ethanol.
9. Remove slides from water tubes.
10. Place slides into tubes containing ethanol and cap.
11. Mix by inverting tubes several times.
12. Remove ethanol, and refill with fresh ethanol.
13. Repeat steps 9–11 for a total of three times.
14. Dry slides under a nitrogen gas stream, or place them on a slide rack and centrifuge to dry.
15. Place dried slides in a sealed container with a desiccant pouch.
16. Store at room temperature until needed.

Key Reference

Matson, R. (unpublished).

Troubleshooting Guide

Note 1: Ethanolamine Reagent-grade ethanolamine should be clear to a light straw color hue. Avoid using reagent that shows a definite yellow coloration or has a strong smell of fish.

Note 2: Ethanolamine Titration Ethanolamine is to be handled with caution. Dilute ethanolamine reagent (pH ~11) is corrosive and requires careful titration with dilute acid. Follow proper laboratory safety: wear goggles, gloves, and a lab coat.

Protocol 3.5: Biotinylated cDNA Target Hybridization to cDNA Slide Microarrays

How It Works

Purified and well-characterized allele-specific cDNAs prepared previously are first immobilized onto a solid support (activated glass slide) in an array format. In order for the tethered strands to participate in hybridization with a complementary target, they must be subjected to a denaturation process. This results in the unwinding of the single strands into sense and antisense probes. The single-stranded probes are then available for annealing to target strands. Similarly, the double-stranded targets (in this case, biotinylated during RT-PCR) must be rendered single stranded prior to hybridization to the solid support. Once both probe and target strands are available, then annealing and hybridization takes place based on thermodynamic principles.

Hybridization typically goes for at least 60 min to overnight at 60°C in a humidified chamber, for example, shaker water bath, depending on the stringency requirements of the experiment.

Preparation time = 20 min
Hybridization time = 60 min (to overnight)
Development time = 100 min
Total time = 180 min

Required Materials

cDNA slide array
Petri dish
Temperature-controlled humidity chamber, for example, shaker water bath
Pipettor
Streptavidin conjugate reporter
Scanner or CCD camera system

Reagent Preparation

Denaturant solution (denaturant) = 0.5 M NaOH, 0.15 M NaCl.
Neutralization solution = 2.4× SSC, 0.016% SDS, 0.28 M Tris, 0.028 M NaCl at pH 7.5, prepared by diluting 1.87 mL of 1.5 M Tris, 0.15 M NaCl, pH 7.5 into 8 mL of 3× SSC, 0.02% SDS.
Hybridization solution = cDNA target mixed into the neutralization solution. The final hybridization solution contains: oligonucleotide target, 2× SSC, 0.01% SDS, 0.24 M Tris, 0.08 M NaCl, ~pH 8.
2× buffer = 2× SSC, 0.01% SDS.

Step-by-Step Protocol

1. Incubate the microarray substrate, for example, glass slide, containing cDNA probes for 15 min, puddle and then spread out uniformly with 150 μL of denaturant solution to produce single-stranded probes. Note: This step is timed to coincide with the end of target denaturation (step 3). The surface of a standard microscope slide may be completely wetted out uniformly with this volume.
2. At the end of the 15 min, rinse the substrate in neutralization solution to neutralize and remove any residual denaturant.
3. Just prior to substrate denaturation (step 1), mix 15 μL biotinylated cDNA with 22.5 μL denaturant in an Eppendorf tube.
4. Incubate for 15 min at room temperature.

5. Next, add 112.5 μL neutralization solution to the tube, and mix well. This becomes the target hybridization (HYB) solution for step 6.
6. Add target HYB solution to the microarray slide.
7. Cover slide, and incubate in a humidified chamber overnight at 60°C.
8. Rinse slide in 2× buffer at 60°C.
9. Rinse slide in 2× buffer at room temperature.
10. Add a suitable streptavidin–dye conjugate (e.g., streptavidin-Cy5) or a streptavidin–enzyme conjugate (e.g., streptavidin–alkaline phosphatase).
11. Develop signal, and read on a suitable laser confocal scanner or CCD camera system.

Key Reference

Matson, R. (unpublished).

Troubleshooting Guide

Note 1: Hybridization

During the hybridization, 2× SSC 0.01% SDS buffer is maintained at 60°C inside 15 mL centrifuge tubes in preparation for the stringency rinse. The development process involves a 10 min stringency rinse in 2× SSC 0.01% SDS inside the 60°C shaker. Slides are placed inside a 15 mL centrifuge tube filled to 10 mL with this stringency rinse buffer. A final rinse in this buffer at room temperature completes the process.

Chapter 4

Protein Sample Preparation for Microarrays

Robert S. Matson

Contents

Introduction

Unlike the preparation of nucleic acid samples, in which only a few chemically distinct classes of molecular species (RNA and DNA) may need to be isolated, the preparation of protein samples for microarray analysis represents a formidable task. It is not that we do not know how to purify proteins. Biochemists have been doing this very well for nearly 100 years, although with this approach, a single protein is methodically isolated to homogeneity and then exhaustively characterized. On the other hand, molecular biology has provided the means to prepare a recombinant protein in sufficient quantities, and with affinity tags that simplify its isolation and subsequent purification. Unfortunately, such processes are best employed in the preparation of standards and solid-phase capture agents such as monoclonal antibodies. In addition, there remains no synthetic approach toward posttranslational modification of nascent proteins. And, although nucleic acids in samples can be mass-amplified and labeled, this has not yet been fully realized for protein samples. The problem with proteins is one of complexity. Simply put, they are structurally diverse and present over a vast concentration range in cells or fluids such as blood. There remains a need to uniformly label the proteins in a sample for microarray analysis. In this chapter, we will introduce approaches to reducing protein sample complexity and means of labeling samples for microarray analysis.

List of Protocols

Protocol Number	Name
4.1	Depletion of Abundant Proteins in Plasma Using IgY Beads
4.2	Trizol Method for Protein Extraction
4.3	Preparation of Protein Lysates from Cultured Cells
4.4	Labeling of Protein Samples with Biotin

Protocol 4.1: Depletion of Abundant Proteins in Plasma Using IgY Beads

How It Works

Plasma represents a very heterogeneous matrix of proteins present in a dynamic range covering over 10 logs in concentration. In order to examine the activity of the estimated 10,000 proteins present in low–medium abundance (protein concentration range, pg/mL–μg/mL) in plasma, it is necessary to remove the most abundant proteins that are present in the mg/mL range. The 12 most abundant proteins in plasma constitute approximately 95% of the total plasma proteome, and with the addition of 10 more proteins, these represent about 98% of the protein mass in plasma. They can swamp out or mask most analyses. Effective depletion of these proteins is accomplished by antibody affinity chromatography; in particular, the use of IgY antibodies raised to serum proteins has been most successful. Avian immunoglobulins (IgY) as antibodies exhibit strong affinity binding toward protein antigens with negligible immuno-cross-reactivity. Furthermore, the IgY molecule is very stable, and easily and rapidly prepared from egg yolk.

The following protocol is based on the use of IgY-bound beads in a spin-column format in which a mixture of beads containing antibodies raised against the 12 most abundant plasma proteins is employed (Figure 4.1). Thus, the high-abundance protein antigens are bound to the beads, thereby depleting the serum. This allows for collection of the unbound fraction, which would then contain the lower-abundance plasma proteins. The bound high-abundance proteins are then released from the immunoaffinity column under acidic conditions and neutralized. Both high- and low-abundance protein fractions can be analyzed using antibody microarrays.

Required Materials

Sample diluent (Tris-Saline): 10 mM Tris, pH 7.4 containing 150 mM NaCl
Elution buffer: 100 mM glycine, pH 2.5
1 M Tris-HCl, pH 8
Spin-filter, 5 kDa molecular weight cutoff (MWCO)
IgY-coupled beads (various mixtures available from GenWay, San Diego, CA); see Note 1 in the section on troubleshooting.
Spin columns (Bio-Rad, Hercules, CA)

Step-by-Step Protocol

1. Dilute 10 μL plasma with 490 μL of sample diluent.
2. Add diluted plasma to the spin column (600 μL bead bed volume) containing the IgY-beads.
3. Cap the column and mix well by end-over-end rotation on a rotator for 15 min.

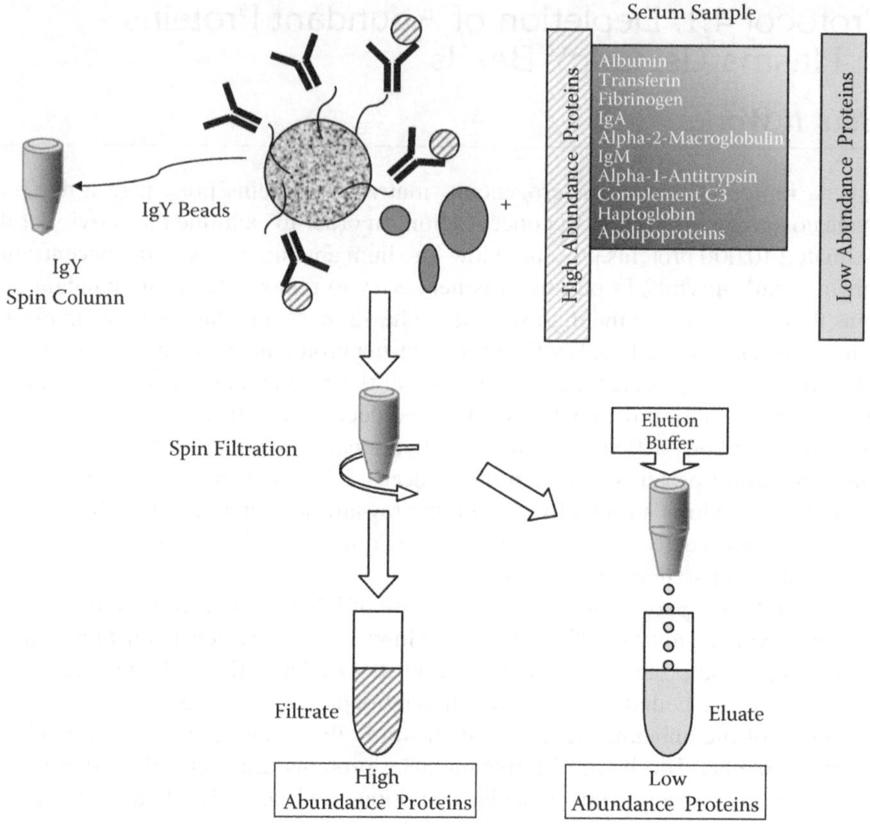

Figure 4.1
Separation of high- and low-abundance serum proteins using IgY spin columns.

4. Remove column from rotator and place in centrifuge.
5. Spin at 400× g for 30 s.
6. Retain filtrate.
7. Apply 500 μL sample diluent to wash the column.
8. Collect column wash #1.
9. Apply an additional volume of diluent to the column.
10. Collect column wash #2.
11. Combine the collected filtrate and the two washes (unbound fractions containing low-abundance proteins).
12. Add 500 μL elution buffer to the column.
13. Cap the column, and mix well by end-over-end rotation on a rotator for 3 min.
14. Remove columns from rotator and place in centrifuge.
15. Spin at 400× g for 30 s.

16. Retain eluate (fraction contains high-abundance proteins).
17. Add 50 µL 1 M Tris-HCl, pH 8 to the eluate in order to neutralize solution.
18. Repeat steps 12–17 for a total of two cycles.
19. Concentrate filtrate (~1.5 mL), and eluate (~1 mL) fractions using MWCO spin filters to approximately 10 µL each for analysis.
20. Store at −20°C until needed; or for long-term storage, aliquot and store at −80°C.

Key References

Anderson, N. L. and Anderson, N. G. (2002). The human plasma proteome: History, character, and diagnostic prospects. *Molecular and Cellular Proteomics,* 1: 845–867.

Desrosiers, R. R. et al. (2007). Proteomic analysis of human plasma proteins by two-dimensional gel electrophoresis and by antibody arrays following depletion of high-abundance proteins. *Cell Biochemistry and Biophysics,* 49(3): 182–195.

Huang, L. et al. (2005). Immunoaffinity separation of plasma proteins by IgY microbeads: meeting the needs of proteomic sample preparation and analysis. *Proteomics,* 5: 3314–3328.

Troubleshooting Guide

Note 1: Custom IgY Beads

If desired, IgY antibodies can be raised, isolated from egg yolk, and purified for coupling to beads. A protocol for the coupling of periodate-oxidized IgY to hydrazide beads is provided in Huang et al. (2005).

Protocol 4.2: Trizol Method for Protein Extraction

How It Works

As described in Protocol 2.2.4, the method involves solution-phase partitioning and extraction of nucleic acids as well as proteins. Tissues are first homogenized in the guanidine isothiocyanate (GuSCN)–phenol reagent. Addition of chloroform results in partitioning of the aqueous phase from the organic phase. RNA remains in the aqueous phase, while DNA and proteins partition into the organic phase. Each of these can be isolated by alcohol precipitation. However, a major drawback of isolating protein has been the often-encountered difficulty in resolubilizing protein for the pellet. Hummon et al. (2007) describe a means of achieving efficient protein solubilization. The protocol described here is adopted from their work. Because the extracted protein is largely denatured, it is best used for the construction of reverse-phase or lysate microarrays (Nishizuka et al., 2003).

Required Materials

TRIzol reagent (Invitrogen, Carlsbad, CA) or TRI reagent (Molecular Research Center, Cincinnati, OH); see Note 1 in section on troubleshooting.
GuSCN–phenol extraction solution (see reagent preparation)
Chloroform
Ethanol
Urea
SDS
Tris-HCl
Eppendorf tubes
Dialysis tubing (Spectra/Por 6; MWCO 2000; Spectrum, Gardena, CA)
Dounce tissue homogenizer (glass–Teflon)
Centrifuge
Pipettor

Reagent Preparation

Guanidinium–phenol extraction solution: Combine 468 mL GuSCN (4 M) with 495 mL phenol. Add 25 g sodium acetate and 1.8 mL 2-mercaptoethanol. Mix well. Adjust to pH 4 with acetate buffer.

Urea–SDS dissolution buffer: Prepare 8 M urea in 50 mM Tris-HCl, pH 8. Add an equal volume of 1% SDS.

Step-by-Step Protocol

1. Place 25 mg tissue into homogenizer, and add 800 μL extraction solution.
2. Form homogenate.
3. Transfer homogenate to Eppendorf tube.
4. Add 80 μL of chloroform to tube, and shake for 15 s to form suspension.
5. Place suspension on ice for 15 min.
6. Centrifuge at 12,000× g for 14 min at 4°C to separate aqueous (upper) from organic (lower) phase.
7. Transfer upper aqueous phase to new tube with pipette (save for RNA isolation).
8. Recover the organic phase.
9. Add an equal volume of ethanol, and mix by vortexing.
10. Place tube in –20°C freezer for 45 min.
11. Centrifuge at 12,000× g for 14 min at 4°C.
12. Observe pellet (save for DNA isolation).
13. Remove supernatant (protein), and transfer into dialysis bag.
14. Dialyze against 0.1% SDS overnight at 2–8°C.
15. Place bags in fresh SDS, and dialyze for 4 h.
16. Place bags in fresh SDS, and dialyze for an additional 2 h.

17. Observe a globular "protein" flocculent precipitate.
18. Carefully remove the retenate solution from the dialysis bag without dislodging the precipitate.
19. Discard the retenate solution.
20. Add Urea–SDS dissolution buffer to dissolve precipitate in dialysis bag.
21. Remove dissolved protein solution, and place in tube for later use.
22. Store at −20°C.

Key References

Chomczynski, P. (1989). U.S. Patent 4,843,155.

Hummon, A. B. et al. (2007). Isolation and solubilization of proteins after TRIzol® extraction of RNA and DNA from patient material following prolonged storage. *BioTechniques,* 42(4): 467–472.

Nishizuka, S. et al. (2003). Proteomic profiling of the NCI-60 cancer cell lines using new high-density reverse-phase lysate microarrays. *Proceedings of the National Academy of Sciences USA,* 100(24): 14229–14234.

TRI Reagent (1995). RNA, DNA, protein isolation reagent. Manufacturer's protocol. Molecular Research Center, Inc., Cincinnati, OH.

Troubleshooting Guide

Note 1: Extraction Reagent

Trizol is the trade name held by Invitrogen (Carlsbad, CA) but is commonly used to describe Chomczynski's reagent. Tri-reagent is the trade name held by Molecular Research Center (Cincinnati, OH) for the extractant.

Protocol 4.3: Preparation of Protein Lysates from Cultured Adherent Cells

How It Works

Cells are grown to confluency and recovered by scraping. A lysis buffer disrupts the cell and suborganelle membranes, spilling out the cytosol into the buffer. Membrane proteins and multimeric protein complexes are denatured and solubilized. The lysates are printed in microarray format.

Required Materials

Cells grown in culture
Cell scraper
Centrifuge

PBS
BCA (Bicinchoninic Acid) protein assay (Pierce, Rockford, IL, Cat. No. 23235)
Protease and phosphatase inhibitor cocktails (Pierce, Rockford, IL; Sigma-Aldrich); protease cocktail, Cat No.78425; phosphatase, Cat. No. 78420; see Note 1 in the following section on troubleshooting
RIPA lysis buffer (Pierce, Rockford, IL, Cat. No. 89900); see Note 2

Reagent Preparation

Lysis buffer: See Note 3.

Step-by-Step Protocol

1. Decant off culture media from T-flask.
2. Wash adherent cells twice with PBS (ice-cold).
3. Remove cells from T-flask using a cell scraper (see Note 4).
4. Transfer to a centrifuge tube in PBS (ice-cold).
5. Pellet the cells by centrifugation at ~200–500× g for 5–10 min.
6. Gently resuspend the cell pellet in ice-cold PBS using a pipette.
7. Pellet the cells by centrifugation at ~200–500× g for 5–10 min.
8. Repeat steps 3–4 an additional time.
9. Determine the total cell count.
10. Resuspend the washed cell pellet in lysis buffer: use 1 mL per ~5×10^6 cells.
11. Add protease–phosphatase cocktails: use ~10 μL per 1 mL lysis buffer volume.
12. Place cell suspension on ice for 30 min for lysis (see Note 5).
13. Centrifuge at 10,000× g for 5–10 min.
14. Recover the supernatant (protein), and transfer to an Eppendorf tube and place on ice.
15. Determine protein content by the Bradford assay or BCA method.
16. Store supernatants at −70°C to −80°C.
17. Preparations for printing lysate microarrays (see Note 6).

Key References

Hummon, A. B. et al. (2007). Isolation and solubilization of proteins after TRIzol extraction of RNA and DNA from patient material following prolonged storage. *BioTechniques,* 42(4): 467–472, 2007.

Jaras, K. et al. (2007). Reverse-phase versus sandwich antibody microarray, technical comparison from a clinical perspective. *Analytical Chemistry,* 79: 5817–5825.

Nishizuka, S. et al. (2003). Proteomic profiling of the NCI-60 cancer cell lines using new high-density reverse-phase lysate microarrays. *Proceedings of the National Academy of Sciences USA,* 100(24): 14229–14234.

Ramaswamy, A. et al. (2005). Application of protein lysate microarrays to molecular marker verification and quantification. *Proteome Science,* 3: 9 [doi: 10.1186/1477-5956-3-9].

Romeo, M. J. et al. (2006). Measuring tissue-based biomarkers by immunochromatography coupled with reverse-phase lysate microarray. *Clinical Cancer Research,* 12(8): 2463–2467.

Sreekumar, A. et al. (2001). Profiling of cancer cells using protein microarrays: Discovery of novel radiation-regulated proteins. *Cancer Research,* 61: 7585–7593.

Winters, M. et al. (2007). Constitution and quantity of lysis buffer alters outcome of reverse phase protein micorarrays. *Proteomics*, 7: 4066–4068.

Troubleshooting Guide

Note 1: Inhibitor Cocktails

Addition of a protease–phosphatase inhibitor cocktail is generally a good idea, to reduce protein degradation during preparation. However, you will need to verify the composition of each cocktail from the vendor. Protease cocktails and phosphatase cocktails are separately formulated, so you may need to add both to the lysis buffer. These are generally supplied in dimethyl sulfoxide (DMSO) to reduce degradation. They are supplied as 10× to 100× stocks so that a small volume (e.g., ~10 μL) can be added to the lysis buffer just before processing. EDTA, which inhibits metalloproteases, is supplied separately because this chelator may interfere in subsequent steps such as metal affinity chromatography or electrophoresis.

Note 2: Use of RIPA Lysis Buffer

RIPA (radioimmunoprecipitation assay) lysis buffer was found to be highly efficient at protein extraction when modified with SDS sample preparation buffer (Winters et al., 2007). This buffer is generally used as a whole-cell protein extractant and does not efficiently remove DNA contaminants. Also, it is inhibitory toward kinases.

Note 3: Lysis Buffers

There are a number of different lysis buffers available for use with this protocol. Most involve a simple buffer such as Tris-HCl, pH ~7–7.5, with a detergent (SDS; Triton) and a reducing agent (DTT; ß-mercaptoethanol). Table 4.1 provides a summary of various lysis buffers used in the preparation of reverse-phase lysate microarrays.

Note 4: Recovery of Cells

Depending on the cell type and culturing conditions, it may be necessary to first trypsinize the cells in the T-flask or culture bottle (see Protocol 2.1.4, Trypsinization to Remove Adherent Cells from Culture Flask). Alternatively, the lysis buffer can be added directly to the attached cells and the lysate collected by aspiration following cell scraping (see Jaras et al., 2007).

Note 5: Sonication

If yields are poor, you can introduce a sonication step to improve solubilization of protein.

TABLE 4.1
Lysis Buffers Used for Protein Extraction

Study	Buffer	Salt	Detergent	Chelator	Inhibitors	Other Additives
Ramaswamy (2005)	50 mM HEPES, pH 7.4	150 mM NaCl	1% TritonX-100	1mM EGTA 1%	10 mM $Na_4P_2O_7$ 100 mM $MgCl_2$ 100 mM NaF	10% Glycerol
Hummon (2007)	1.25 mM Tris-HCl, pH 7.5	150 mM NaCl	0.01% NP40	8 μM EDTA	5 mM $Na_4P_2O_7$ 5 mM NaF 0.1 mM Na_3VO_4	
Nishizuka (2003)	2% Pharmalyte, pH 8.0–10.5		4% CHAPS			9 M Urea 65 mM DTT
Romeo (2006) T-PER Buffer (Pierce)	25 mM Bicine, pH 7.6	150 mM NaCl	Proprietary (Pierce) + 1X SDS Sample Buffer (Invitrogen)			2% ß-mercaptoethanol
Sreekumar (2001)	PBS, pH7.2		1% NP40		Protease Inhibitor (complete, EDTA-free, Roche Diagnostics)	
Winters (2007) RIPA (Pierce)	25 mM Tris-HCl, pH 7.6	150 mM NaCl	1% NP40 0.1% SDS			1% sodium deoxycholate

Note: SDS sample buffer: 2% SDS + 9% glycerol.

Note 6: Preparation for Printing

It may be necessary to take steps that will further denature and solubilize the lysate for printing. To do this, prepare the sample in, for example, 1% SDS + 1% ß-mercaptoethanol or Laemmli sample buffer used for denaturing electrophoresis, and then heat at 94°C for 5 min or boiling water for 2 min. Alternative, the sample may be prepared using the Urea–SDS dissolution buffer described previously in Protocol 4.2.

Protocol 4.4: Labeling of Protein Samples with Biotin

How It Works

Antibody microarrays in which analyte-specific capture antibodies are immobilized do not present a significant challenge when it comes to labeling, detection, and analysis. Nevertheless, there are certain limitations in working with multiplexed assays, especially regarding the issue of cross-reactivity and the matching of antibody pairs. However, the basic processes remain those of the well-established sandwich immunoassay. This is not the case for the reverse-phase lysate microarray. This format requires labeling of all the proteins present in, for example, the serum sample to interrogate the proteins immobilized in the lysate. The problem encountered with this approach is how to achieve uniform and consistent labeling of a large protein population. Direct labeling with dyes such as Cy3/5-NHS is an approach that has been used to compare two different sample protein populations by the differential analysis format originally designed for gene expression microarrays. There are, however, problems associated with the degree of dye incorporation into different proteins for different dyes. Another approach is to first label with biotin and then to postlabel with a streptavidin reporter such as streptavidin-Alexa or –Cy3/5, alkaline phosphatase, etc. Biotinylation of most proteins is straightforward and efficient. There are a variety of biotinylation reagents to choose from that offer different spacer arms and reactive groups which can be coupled to proteins. The most common reagent is the bifunctional reagent NHS-Biotin, which has biotin terminated with the active NHS ester. If a spacer arm is inserted between the biotin and the ester, then this is designated by NHS-XX-Biotin, or NHS-LC-Biotin, where LC refers to a long-chain (LC) alkyl group as a spacer.

Required Materials

Serum protein sample (65–85 mg/mL, total protein)
Biotinylation reagent, e.g., EZ-Link Sulfo-NHS-Biotin (Pierce, Rockford, IL, Cat. No. 21425)
PBS
Centrifuge
Microcentrifuge tube
Pipettor
Desalting column (open column) or spin column, e.g., Zeba™ Desalt Spin Column, 7 kDa MWCO (Pierce, Rockford, IL, Cat. No. 89891)
Dialysis tubing, e.g., 3.5 kDa MWCO (Spectrum Laboratories, Hercules, CA)

Reagent Preparation

Sulfo-NHS-Biotin: Immediately prior to use, dissolve 1.1 mg in 250 µL DDI water. This is enough to prepare 5 samples; see Note 1 in the following section on troubleshooting.

Step-by-Step Protocol

1. Clarify serum sample by centrifugation.
2. Carefully aspirate a small volume (20 µL) of the supernatant using a pipettor (see Note 2).
3. Dilute serum supernatant in PBS to approximately 1 mg/mL in a 1.5–2 mL microcentrifuge tube.
4. Add 25 µL of 10 mM Sulfo-NHS-Biotin to 1 mL diluted serum (see Note 3).
5. Cap tube and place on end-over-end rotator.
6. Incubate for 1 h at room temperature.
7. Transfer solution into a desalting column previously equilibrated in PBS.
8. Collect the protein fraction for use or store at –20°C until needed.

Key References

EZ-Link Sulfo-NHS-Biotinyation Kit Instructions (Pierce, Rockford, IL, Cat. No. 21425).
Ingvarsson, J. et al. (2007). Design of recombinant antibody microarrays for serum protein profiling: targeting of complement proteins. *Journal of Proteome Research,* 6: 3527–3536.

Troubleshooting Guide

Note 1: Preparing Diluted Serum Protein

Assume an average protein concentration of 75 mg/mL. Therefore, 15 μL contains ~1 mg total serum protein. Dilute 15 μL with 985 μL PBS to achieve 1 mg/mL.

Note 2: Handling Sulfo-NHS-Biotin

Sulfo-NHS-Biotin is moisture sensitive. Bring the reagent to room temperature before opening. The NHS ester will undergo hydrolysis in aqueous solutions, and should be prepared and used within a few minutes.

Note 3: Determination of Biotin Molar Excess

To estimate molar excess, one needs to determine the number of moles of protein. Serum protein comprises ~60% albumin (66.4 kDa) and ~35% immunoglobulin (mostly IgG, 150 kDa), so odds are that you will be mostly biotinylating albumin and immunoglobulin. One can use a weighted average of these two proteins for the purpose of calculation. Therefore, 1 mg/mL serum protein would be 11.4 nmol in 1 mL. A 20-fold excess of NHS-Biotin (mol. wt. = 444.3) would require $20 \times 11.4 =$ 228 nmoles of reagent. If the NHS-Biotin concentration is 10 mM (10 nmol/μL), then 22.8 μL would be needed, so use 25 μL.

Chapter 5

Solid-Phase Chemistries for Protein Microarrays

Robert S. Matson

Contents

Introduction

Protein Immobilization Chemistries

Proteins are arrayed as capture agents onto planar substrates. Depending on the properties of the solid-phase substrate and the protein, this can be accomplished either by adsorption (noncovalent) or covalent attachment. Proteins can be adsorbed based on hydrophobic and electrostatic interaction with the substrate. In this case, it is important to find conditions that avoid surface denaturation. Covalent attachment is generally accomplished via the linking of amine-reactive groups on the substrate to exposed lysine residues of the protein. Most work to date has been with antibodies that are arrayed and then used to capture antigens that are subsequently detected using labeled secondary antibodies. This is commonly referred to as a *sandwich immunoassay*. Thus, an antibody array is a multiplexed, miniaturized sandwich immunoassay.

List of Protocols

Protocol Number	Name
5.1	Passive Adsorption
5.2	Covalent Attachments

Protocol 5.1: Passive Adsorption

How It Works

Using the following protocol, antibodies are passively adsorbed onto a glass slide coated with poly-L-lysine.

Protocol 5.1.1: Immobilization to Poly-L-Lysine Slides

Required Materials

Arrayer
384-well microplate
Poly-L-lysine-coated microscope slides
PBS buffer, pH 7.4
Nonfat milk (blocking agent)
Tween 20
Sodium azide (preservative)
Dialysis tubing

Reagent Preparation

PBST: Add 100 μL Tween 20 to 100 mL of 10 mM PBS, pH 7.4, and mix well.
3% nonfat milk: Add 3 g of milk to 100 mL PBST.

Step-by-Step Protocol

1. Transfer antibodies from glycerol buffer into glycerol-free PBS solution by dialysis (see Note 1).
2. Prepare proteins at 0.1–0.3 mg/mL.
3. Transfer 10 μL into 384-well source plate (see Note 2).
4. Array solutions from source plate onto the poly-L-lysine-coated glass slide.
5. Rinse microarrays briefly in 3% nonfat milk in PBS, pH 7.4, containing 0.1% Tween 20 (see Note 3).
6. Soak slides overnight at 4°C in 3% nonfat milk in PBS, pH 7.4, containing 0.02% sodium azide as preservative.
7. Rinse slides at room temperature with PBS just prior to use.
8. Maintain slides in PBS buffer until incubation with sample.

Key Reference

Haab, B. B. et al. (2001). Protein microarrays for highly parallel detection and quantitation of specific proteins and antibodies in complex solutions. *Genome Biology*, 2(2): research0004.1–0004.13.

Troubleshooting Guide

Note 1: Slides Protocol developed for a poly-L-lysine slide only. Glycerol causes spreading and may reduce binding to substrate. Use spin columns or microdialysis for smaller sample sizes for buffer exchanges to remove glycerol.

Note 2: Source Plate Generally, a 384-well microplate is used as a source plate for the arrayer system rather than a 96-well plate. The 384-well plate permits the use of greatly reduced sample volumes, the arraying of larger numbers of samples, and is less prone to evaporation if properly stored. However, this really depends on the needs of the user, the arrayer setup, and the number of slides that are to be printed.

Note 3: Buffers Nonfat milk or casein blocks nonspecific sites on slide, whereas Tween 20 serves two purposes: (1) Removal of nonspecific protein and (2) hydration of bound protein reducing denaturation. It is therefore important to maintain hydration. Do not allow the slides to dry out during processing.

Protocol 5.2: Covalent Attachments

How It Works

Covalent attachment is generally accomplished via the linking of amine-reactive groups on the substrate to exposed lysine residues of the protein. For protein microarrays, the most commonly employed chemistries are aldehyde and epoxide surfaces. For efficient coupling, the reaction is carried out under basic conditions (pH 8.5–10) such that the lysine residues are not fully protonated, ensuring that the unshared pair of electrons on the nitrogen are freely available to participate in the nucleophilic attack. In the case of the aldehyde reaction, a Schiff's base is formed. Depending on the number of lysine residues participating in the coupling, it may be necessary to reduce the Schiff's base by borohydride reaction. The reduction of the Schiff's base (which is not a formal covalent bond) results in conversion to the alkyl amine, which represents a formal and rather strong covalent bond.

In Protocol 5.2.1 (described next), agarose is first converted by a periodate oxidation reaction to form aldehyde-agarose. Subsequently, the amine-reactive protein is immobilized by the formation of the Schiff's base between the aldehyde groups on the agarose and the lysine residues on the protein. The Schiff's base is reduced by the borohydride reaction (Figure 5.1).

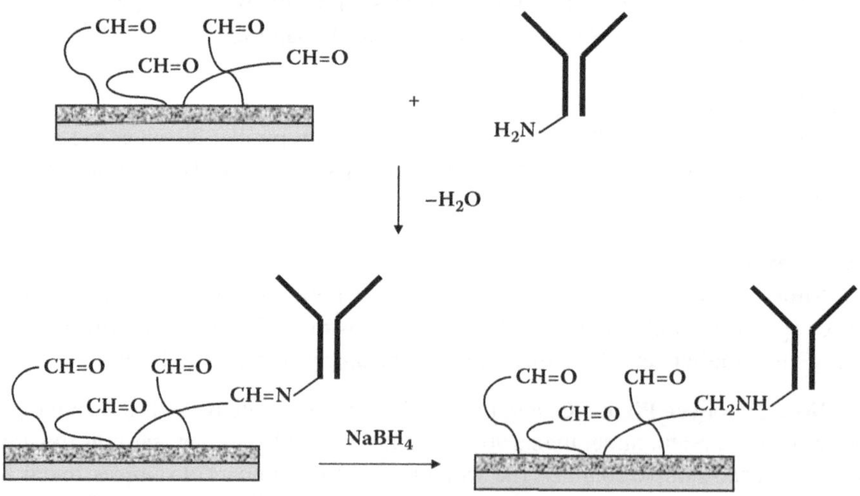

Figure 5.1
Attachment of antibodies to aldehyde-agarose via Schiff's base.

Protocol 5.2.1: Preparation of Amine-Reactive Substrates Based on Aldehydes

Required Materials

Aminopropyl-silanized microscope slides (see Note 1).
Agarose
Ethanol
PBS
PBS–Tween 20 buffer (PBST)
SDS, 0.2%
Sodium bicarbonate buffer, 50 mM, pH 9.6
Sodium borohydride ($NaBH_4$), FW 37.83
Sodium m-periodate ($NaIO_4$), FW 213.89
384-well microplate
Antibodies
Dialysis tubing
Humidity chamber

Reagent Preparation

Sodium m-periodate (20 mM) solution: Dissolve 214 mg in 50 mL DDI water.
Sodium borohydride (50 mM) solution: Dissolve 95 mg in 40 mL PBS + 10 mL ethanol.
PBST: Add 100 µL Tween 20 to 100 mL of 10 mM PBS, pH 7.4, and mix well.

Step-by-Step Protocol

1. Prepare agarose (1% v/v) by dissolving 1 g agarose in 99 mL DDI water.
2. Warm the agarose solution to about 70°C.
3. Pipette 2 mL of the agarose over the microscope slide and allow to gel.
4. Air-dry the slides.
5. Fill a 50 mL conical polypropylene centrifuge tube with 40 mL of 20 mM sodium m-periodate solution (see Note 2).
6. Soak slides (two per tube, back to back) in the conical tube for 30 min at ambient temperature.
7. Remove the solution, and carefully add distilled water without disturbing the agarose surface.
8. Decant off the distilled water, and replace with fresh distilled water.
9. Soak for 2–3 h, then decant.
10. Air-dry the slides and store in a dry container.
11. Prepare antibodies at 1 mg/mL in 50 mM sodium bicarbonate buffer, pH 9.6, by dialysis (see Note 3).
12. Transfer dialyzed proteins to a 384-well source plate for printing.
13. Print from the source plate onto the aldehyde-activated agarose slide (step 10).
14. Place the arrayed slide in a humidified box, and incubate overnight at 4°C.
15. Soak the slide at room temperature by submerging in 50 mL conical filled with PBST for 5 min, followed by two rinses in the same buffer.

16. Transfer the slide into a new 50 mL conical containing 40 mL of sodium borohydride solution, and soak the slide for 5 min. (see Note 4).
17 Remove the slides and transfer into a new conical tube for washing.
18. Wash the slides sequentially with the following using a new conical at each step:
    a. 0.2% SDS—wash three times, each with a 2 min soak.
    b. DDI water—wash two times, each with a 1 min soak.
19. The antibody array is now ready for use. Otherwise, store dry under refrigeration (4°C–8°C).

Key Reference

Afanassiev, V. et al. (2000). Preparation of DNA and protein micro arrays on glass slides coated with an agarose film. *Nucleic Acid Research,* 28(12): e66.

Troubleshooting Guide

Note 1: Slides It is important that the microscope slides be clean prior to the addition of the agarose. Soak slides in a mild detergent for 10 min, and then rinse extensively using DDI water. Rinse off with reagent-grade methanol or ethanol. Allow to air-dry. Do not handle without wearing gloves. Avoid blotting with tissue, because this may introduce lint onto the slides.

Note 2: Periodate Solution This solution must be prepared just prior to use.

Note 3: Dialysis Antibodies obtained from vendors may have stabilizing agents such as glycerol and bovine serum albumin (BSA) (1%) that should be removed prior to use. The easiest way to accomplish this is by dialysis against the coupling buffer, which also equilibrates the sample. If many samples are needed, then spin columns can be used to exchange buffers.

Note 4: Borohydride Solution This solution must be prepared just prior to use. It is good for about 1 h. The conical tube cap should remain loose to avoid pressure buildup from the borohydride reduction, which produces hydrogen gas as a by-product of the reaction. As the pH of the solution turns acidic from hydrogen production, the borohydride decomposes. Ethanol is added to reduce bubble formation. To get rid of excess reagent, increase to pH 4 with dilute acid and set the solution in the fume hood overnight.

Protocol 5.2.2: Immobilization by Covalent Attachment to Surface Epoxides

How It Works

In the following protocol, antibodies are covalently attached to a glass slide having an amine-reactive surface. In this case, the slide's surface contains epoxide groups that will react with the lysine residues found on proteins under basic pH conditions.

Required Materials

Epoxide slide
Hammerstein casein (blocking agent)
Sodium carbonate
Sodium bicarbonate
PBS
NaCl
NaOH (1 M)
Tween 20
Sodium azide (preservative)
Dialysis tubing
384-well microplate
Humidity chamber

Reagent Preparation

Print buffer: sodium borate, 100 mM, pH 9.

Carbonate buffer: 100 mM sodium carbonate buffer, pH 9.8, containing 150 mM NaCl.

Casein quench-block solution: Dissolve 10 g of Hammerstein casein in 100 mM sodium carbonate buffer, pH 9.8, by gentle stirring. Adjust pH with 1 M sodium hydroxide and bring volume up to 100 mL to achieve 50 mM buffer and 10% casein with distilled water.

PBST rinse buffer: Add 100 µL Tween 20 to 100 mL of 10 mM PBS, pH 7.4, and mix well.

Step-by-Step Protocol

1. Prepare antibodies in amine-free and thiol-free buffers by dialysis into PBS, pH 7.4, at 1–2 mg/mL protein.
2. Dilute the dialyzed antibodies into print buffer at 0.2–1 mg/mL final concentration (see Note 1).
3. Transfer 60 µL of the diluted antibody solution into 384-well source plate for printing.
4. Array solutions from source plate onto the epoxide-coated glass slide.
5. Place arrayed slide in a humidity box, and incubate for 1 h at room temperature.
6. Transfer slide into 50 mL conical tube filled with 40 mL of casein quench-block solution, lay the tube on its side on a shaker, and shake for 1 h at room temperature.
7. Remove the casein solution and replace with 40 mL PBST rinse buffer. Incubate 5 min with gentle shaking.
8. Discard rinse buffer, and repeat step 7 twice more.
9. Rinse slide in DDI water, and air-dry.
10. Store slide in a dry container under refrigeration until ready for use.

Key Reference

Agenendt, P. et al. (2003). Next generation of protein microarray support materials: evaluation of protein and antibody microarray applications. *Journal of Chromatography A*, 1009: 97–104.

Troubleshooting Guide

Note 1: Protein Concentration It has been our experience that printing arrays at an antibody protein concentration of ~200 µg/mL is sufficient for capture in a sandwich assay. However, capture efficiency is dependent on the nature of the immobilized antibody and, in particular, the resulting antigen affinity, which may be different from that reported by the vendor. It is therefore recommended that the user first undertake a series of studies aimed at optimizing antibody surface loading.

Note 2: Blocking Buffers It may be necessary for the user to substitute for other blocking buffers (buffer, pH, proteins, and other additives), depending on the nature of the assay under development by the user.

Protocol 5.2.3: Immobilization by Covalent Attachment to Hydrogel Slide Epoxides

How It Works

Antibodies are arrayed as capture agents onto three-dimensional hydrogel substrates. Although most protein microarray studies have involved the use of planar slides, the use of polymer (hydrogel)-coated slides offers the advantages of increased binding capacity and greater sensitivity. This is accomplished essentially by increased surface area for immobilization of the antibody and by a reduction in steric hindrance. In the protocol described in the following text, the sugar, trehalose, is included in the print buffer. Trehalose is known to preserve protein functionality in the dry state, such as during lyophilization, by keeping the protein hydrated.

Required Materials

Hydrogel epoxide slide
Trehalose
Polyvinyl alcohol
PBS
Dialysis tubing
384-well microplate
Arrayer
Humidity box or suitable container

Reagent Preparation

Print buffer: 15 mM sodium phosphate buffer, pH 7.2, containing trehalose (40% w/v)
Blocking buffer: PBS, pH 7.2, containing polyvinyl alcohol (1% w/v)

Step-by-Step Protocol

1. Prepare antibodies in amine-free and thiol-free buffers by dialysis into PBS, pH 7.4, at 1–2 mg/mL protein.
2. Dilute the dialyzed antibodies (1:10 v/v) into print buffer at 0.1–0.2 mg/mL final concentration.
3. Transfer 60 μL of the diluted antibody solution into 384-well source plate for printing.
4. Array solutions from source plate onto the epoxide-coated glass slide.

5. Place arrayed slide at 4°C overnight in a humidity box (see Note 1).
6. Transfer slide into 50 mL conical tube filled with 40 mL of blocking buffer, lay the tube on its side on a shaker, and shake for 40 min at room temperature.
7. Wash slide in distilled water.

Key Reference

Zubtsov, E. N. et al. (2007). Comparison of surface and hydrogel-based protein microchips. *Analytical Biochemistry*, 368: 205–213.

Troubleshooting Guide

Note 1: Hydrogels It is important to maintain hydration of the slide at all times. Otherwise, drying of the hydrogel may lead to irreversible damage or alter its adsorptive nature and binding properties.

Chapter 6

Protein Microarrays

The Link between Genomics and Proteomics

Persis P. Wadia and David B. Miklos

Contents

Introduction

In the early 1980s, Ekins's "Ambient Analyte Immunoassays" manuscript described the basic principles of microarray technology; it stated that minimal amounts of binding materials remain at the same concentration even in small amounts of the sample, and that these minute amounts of binding material can give better detection sensitivity.[1] Also, cDNA cloning and gene identification increased rapidly in the late 1980s, driving a desire to analyze hundreds of gene or gene products concurrently. This resulted in miniaturized and parallel measurements of large number of targets based on the use of high-density microarray technology. Since 1987, the use of multiplexed assessment of gene expression using cDNA microarrays or DNA oligonucleotide chips has grown exponentially, and it now facilitates gene expression profiles across organisms from bacteria to humans.[2] In line with the central dogma of biology that protein follows DNA and RNA synthesis, the multiplexed assessment of protein function and immune recognition has lagged behind. Although the high-density protein microarray remains a nascent technology, researchers are developing new applications as content expands.

Traditionally, proteins were characterized on the basis of their enzymatic activities such as glucose metabolism through biochemical properties and chromatography separation.[3] One- or two-dimensional gel electrophoresis[4] provided rapid separation of complex protein mixtures, and the presence of a single protein could be determined by specific antibody recognition on a Western blot. Enzyme-linked immunosorbent assays (ELISAs) enabled rapid quantification of isolated single proteins in solution. However, these techniques remained labor intensive, with only a few proteins being detected in each experiment and their identification usually requiring protein-specific antibodies. Mass spectrometry (MS) has become the rapid method for lysate protein profiling when coupled with matrix-assisted laser desorption–ionization (MALDI) or surface-enhanced laser desorption–ionization (SELDI). These techniques are particularly useful when screening for proteins between two sample sets with only small amounts of starting material or in the identification of an unknown purified protein.

These methods were designed to discover individual proteins. However, now that the complete gene composition has been determined[5] and protein composition can be inferred from gene expression profiles, research requires multiplexed methods to sample enzymatic activity, direct protein binding, or antibody recognition across the entire proteome and frequent polymorphic variants.

High-density protein microarray technology meets the aforementioned research need. Entire proteomes can be recombinantly expressed, affinity-purified, and immobilized on microarrays. These can be probed with fluorescent- or radionucleotide-labeled substrate, protein, DNA/RNA, or antibody. Alternatively, primary protein or antibody binding can be detected by secondary fluorochrome conjugated antibodies.

Because microarrays inherently test hundreds to thousands of targets concurrently, the most useful microarrays include representation of the entire proteome of the organism. This has been achieved in simpler eukaryotes such as *Saccharomyces cerevisiae*,[6] but commercially available human microarrays are currently limited to 8000 unique proteins or about 25% of the proteome. Quality assurance and normalization requires a variety of repetitively printed positive and negative controls, geographically distributed throughout the microarray.[7] Microarrays printed with proteins recombinantly expressed may incorporate epitope tags (6-histidine or small peptide epitopes, e.g., V-5 epitope), facilitating affinity purification. These can be detected by monoclonal antibody recognition after microarray printing in order to confirm matrix binding and facilitate printed antigen normalization. Purified fusion protein alone or epitope-tagged negative control proteins can be printed, with increasing known concentrations providing reference for protein printing/binding normalization. This epitope-tag normalization strategy is limited by variable epitope presentation in the final recombinant protein structure.

Two kinds of protein arrays can be designed: the forward-phase and reverse-phase arrays. Forward-phase arrays are those in which an array of well-defined capture molecules are printed on the array, and thus, in one application of the sample, multiple analyses of numerous parameters can be achieved.[8] Examples are antibody arrays for different cytokines; however, one can also study protein–protein/protein–DNA interactions, protein–drug interactions, enzyme–substrate interactions, etc. Reverse-phase arrays are those in which different samples, for example, tissue or cell lysates, are printed. Hence, each printed spot contains the whole repertoire of protein profile of a cell or tissue, which can be screened for the presence or absence of one target protein.[9] An example is screening for biomarkers in a set of tissue arrays specific to a region, for example, skin affected with acute graft-versus-host disease (GVHD) as compared to a set of tissue arrays from skin not affected with GVHD.

As a practical matter, protein microarrays are typically glass slides coated with a variety of substrates such as nitrocellulose, polyvinylidene fluoride (PVDF), agarose, cellulose, nylon, or polyacrylamide.[10] These substrates need important modifications

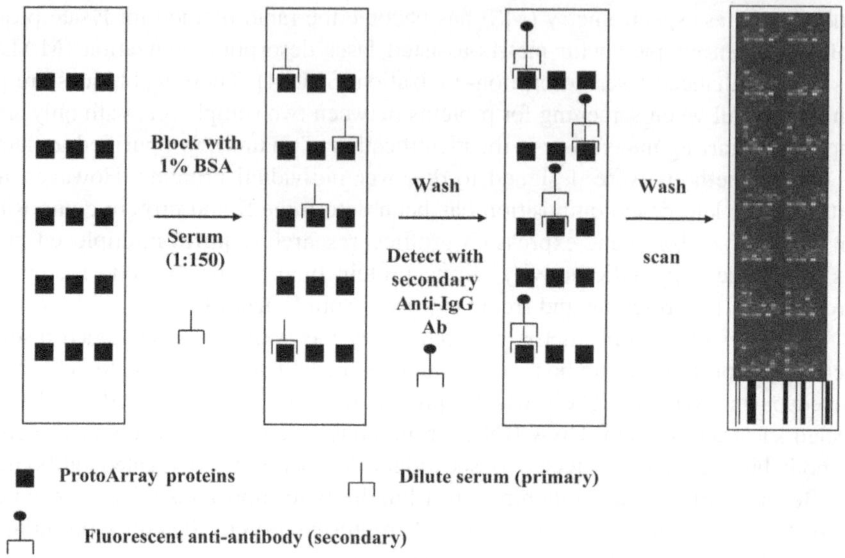

Figure 6.1
Method for processing a protein microarray. Proteins of interest are printed on a nitrocellulose-coated glass slide. Primary antibody is applied in the form of diluted sera/plasma followed by the addition of a secondary antibody conjugated to a fluorochrome. The slides are washed, dried, and scanned at the respective wavelength.

as glass surfaces show poor protein binding capacity. Proteins can be attached using affinity binding for, for example, antibodies; or streptavidin for biotinylated proteins; or nickel for histidine-tagged proteins. Each spot measurement can be a nanometer or a millimeter depending on the size of the array or the number of spots on it. Proteins can be deposited through the use of either contact or noncontact dispenser technologies.[11] For contact printing, arrayers have needles or contact pins that transfer the protein solution from a microwell plate onto the substrate. Noncontact deposition arrayers use capillaries to deposit the protein. These arrayers usually have 96 deposition capillaries and deposit 96 proteins at once as a spray using a piezoactuator.

Once the experiment has been performed with the test and control samples, software packages are available to analyze the data generated. These packages include, among others, GenePix™ (MDS Analytical Technologies),[12] SCANARRAY™ (PerkinElmer), and ImageQuant™ (GE Healthcare).[9] Once the data are scanned and processed by the software, care should be taken to reanalyze each spot with its respective values. This can also be achieved by printing headers in the same location in each array or subarray for alignment purposes (Figure 6.1). Protein microarrays can also be bought commercially. These protein microarrays from Invitrogen (www.invitrogen.com/protoarray) display 8000 full-length human proteins with N-terminal GST epitopes expressed in baculovirus infected insect cells and affinity purified under native conditions maintaining their cellular enzymatic activities and native conformations.

Concerns regarding protein microarrays include the following:

1. The reactivity of proteins and their binding sites should be preserved through purification, microarray binding, and storage.
2. Correct orientation of the proteins should be obtained so that the immunogenic epitopes are exposed.
3. Analysis methods and bioinformatics have not been standardized, and standard operating procedures (SOPs) should be maintained for analyzing all test and control samples.
4. Normalized protein or antigen should be loaded on each array for consistent and reproducible results. Evaporation of content during printing time needs to be tested and controlled.
5. Protein microarrays to date are limited with respect to the number of different analytes that can be used as compared to nucleic acid arrays.
6. Posttranslational modifications such as glycosylation or phosphorylation may affect protein–protein interactions depending on recombinant expression system.
7. Protein array density remains limited compared to DNA microarray chips; thus, results differ depending on method and hypothesis.

Our laboratory's main microarray application involves plasma antibody target identification. The major technical challenges include varying antigen printing/binding concentration, intersample normalization, and inferring antibody binding strength/concentration (titer) from single plasma dilution microarray measurements. The ability to print antigen at high concentration is critical in meeting the dynamic range for plasma antibody that allows broad antibody concentrations to be compared. In addition, monoclonal antibody detection and normalization of printing/binding for all antigens facilitate analysis. Of course, printing replicates determine experimental variation.

This is a rapidly evolving technology that can be applied to various diseases, such as types of cancers, to identify biomarkers and can be easily integrated with gene expression or single-nucleotide polymorphism (SNP) data in the study of complex diseases, including autoimmunity and cancer. Serum/plasma profiling from various viral/fungal infections or from various cancers can lead to new biomarkers that, when validated, can be used by clinicians to provide early diagnosis and also by scientists in drug discovery.

In this chapter, we will describe our method for identifying human protein targets of new antibodies that develop after an allogeneic hematopoietic cell transplant (HCT). Our work is based on the use of commercially available human protein microarrays as well as our experience in printing custom protein and peptide microarrays.

Scientific Background

Allogeneic HCT cures some hematologic cancers as a result of the donor's immune system recognizing the patient's cancer as foreign and immunologically destroying the cancer. Certainly, T-cells play key roles in posttransplant alloimmune responses, but T-cell epitope determination remains laborious, being cell culture dependent and HLA restricted.[13] However, there is growing evidence that co-coordinated humoral alloimmunity develops after HCT and associates with important clinical outcomes.[14] New antibodies develop after allogeneic HCT against minor histocompatibility antigens encoded on the Y chromosome, called *H-Y antigens*. In patients undergoing sex-mismatched HCT, there is association of H-Y antibody development with both chronic GVHD development and persistent disease remission.[15] We hypothesize that disease-specific minor histocompatibility antigens (mHA) are targeted by new antibodies detectable one year after allogeneic HCT, using protein microarray technology in comparison with respective pretransplant sera results. Commercially available (ProtoArray™—Invitrogen, Carlsbad, CA) microarrays present 8000 human proteins enabling high-throughput, quantifiable serologic screening for targets of allogeneic B-cell responses, that is, allo-Ab. Patient plasma collected before and after HCT can be probed using these arrays to determine a profile of allo-Ab targets in an unbiased manner. Moreover, microarray content promises to include the entire human proteome soon. This "broad-band" technology therefore facilitates high-throughput genomic characterization supporting a human systems biology approach to HCT termed *immunogenomics*.

We have shown that allo-Abs develop in patients undergoing sex-mismatched allogeneic HCT in association with both chronic GVHD development and persistent disease remission.[15] A major research goal is to understand epitope preference, epitope spread or restriction over time, and isotype switching. All require concurrent evaluation of antibody development against hundreds of protein and peptide epitopes. This has been made possible using microarray technology. This chapter will describe the development and printing of custom protein and peptide arrays, and the exploratory use of commercially available human protein microarrays to discover the targets of new antibody responses after allogeneic HCT.

List of Protocols

Protocol Number	Name
6.1	Sample Sources and Their Preparation
6.2	Preparation and Characterization of Antigens
6.3	Printing of Protein Microarrays
6.4	Assay Development
6.5	Storage

Protocol 6.1: Sample Sources and Their Preparation

How It Works

Serum or plasma can be used for protein microarrays. Usually, plasma is better than serum for discovery assays for biomarkers because fibrin clotting could remove relevant biomarkers from sera. Blood is collected in tubes containing sodium heparin for plasma. Fibrinogen or albumin in the plasma or debris/floccules in the serum/plasma samples should be removed before applying the sample on the protein arrays (see Note 1 in the section on troubleshooting). This is an important initial step in the processing of protein microarrays.

Required Materials

Venous blood collection tubes (BD Vacutainer®, Franklin Lakes, NJ, Cat. No. 367874 or BD Vacutainer serum tubes)
21G $^{3}/_{4}$ Butterfly blood collection needles (BD Vacutainer, Franklin Lakes, NJ, Cat. No. 367251)
Needle holders (BD Vacutainer, Franklin Lakes, NJ, Cat. No. 364893)
Tourniquet (BD Vacutainer, Franklin Lakes, NJ)
Alcohol swab (VWR, West Chester, PA, Cat. No. 56612-916)
Needle disposal container for needles used to collect blood
Cryovials (Fisher Scientific, Pittsburgh, PA, Cat. No. 0566966)
Centrifuge (Beckman Coulter, Fullerton, CA, Model: Allegra 6 Centrifuge)

Step-by-Step Protocol

Plasma Collection

1. Label tubes with Patient/Control ID. For plasma, BD collection tubes (Cat. No. 367874) are used, and if serum is required, then BD serum tubes are used.
2. The site of the vein to be punctured is cleaned with an alcohol swab.
3. Place the tourniquet above the site selected for venipuncture, and venipuncture with an 18–22 gauge phlebotomy needle until venous blood flash is seen above needle shaft.
4. The needle is either directly or via butterfly tubing set connected to vacutainer sample tubes or filled to the predetermined level (5–10 mL) to adequately dilute heparin anticoagulant. Blood flow stops once vacutainer-tube vacuum is depleted by blood.
5. Remove the tourniquet before the needle is withdrawn.
6. Invert the tube gently five times, and apply pressure to the venipuncture site with an appropriate bandage.
7. Centrifuge the tubes at 2500–3000 rpm for 10 min at 4°C within 2 h of blood collection, transfer 100 uL to 2 mL plasma aliquots into cryopreservation tubes within 1–2 h, and store at either −20°C or −80°C.

Preparation of Plasma/Serum for Use on Protein Microarrays

1. Thaw frozen aliquots on ice, and if repetitive use is anticipated, prepare smaller aliquots to minimize future freezing/thawing.
2. Cryopreserved plasma develops a *cryoprecipitate* composed predominately of von Willebrand factor and Factor VIII clotting factors. This globular material can be removed from the thawed plasma using a clean 200 μL pipette and discarded. The remaining cryoprecipitate is pelleted by microcentrifugation at 14,000 rpm for 10 min at 4°C.
3. Necessary plasma is removed from the top of the tube for further testing.

Protocol 6.2: Preparation and Characterization of Antigens

How It Works

While purified recombinant proteins can be expressed from a myriad of organisms, we predominately express epitope-tagged human proteins from either insect cells or in bacteria (*Escherichia coli*). Affinity purification and normalization requires the incorporation of an epitope tag (His, V5, GST, etc.). Epitope tagging facilitates microarray quality assessment. Printing and binding efficiency can be directly assessed for all recombinant proteins by probing with a tag-specific monoclonal antibody. Protein concentration for print application can be adjusted using epitope tag detection to normalize the printing of recombinant proteins. A protocol for a protein microarray with custom proteins is described in the following sections. These proteins are GST-tagged and also have a 6-histidine tag.

Required Materials

LB agar with 50 μg/mL Ampicillin plates (Teknova, Hollister, CA, No. L1150; VWR Cat. No. 100216-734)
2XYT broth (EMD Chemicals, Inc. (BIOSCIENCES, Gibbstown, NJ, Cat. No. 71755-3)
1 g Ampicillin (BD Diagnostics, Burlington, NC, Cat. No. 230705)
IPTG (EMD Chemicals, Inc. (BIOSCIENCES, Gibbstown, NJ, Cat. No. 5810)
B-PER Bacterial Protein Extraction Reagent (Pierce Chemical, Rockford, IL)
NuPAGE® Sample reducing agent (10×) (Invitrogen, Carlsbad, CA, Cat. No. NP0004)
NuPAGE® MOPS SDS running buffer (20×) (Invitrogen, Carlsbad, CA, Cat. No. NP0001)
NuPAGE® LDS 4× LDS sample buffer (Invitrogen, Carlsbad, CA, Cat. No. NP0008)
NuPAGE® Novex 4–12% Bis-Tris gel 1.0 mm, 12 well (Invitrogen, Carlsbad, CA, Cat. No. NP0322BOX)
Anti-GST antibody (Upstate, Cat. No. 05-311)
Secondary antibody: Peroxidase Affini Goat Anti-Human IgG, Fc_γ Fragment Specific (min x Bov, Hrs, Ms Sr Prot) Jackson Laboratories, Inc., Cat. No. 109–035–098
TBS buffer (Invitrogen, Carlsbad, CA, Cat. No. R017R.0000)
1.5 mL Eppendorf tube (Eppendorf, San Diego, CA, Cat. No. 022364111)
Eppendorf centrifuges (Eppendorf, San Diego, CA, Cat. No. 5417 R)
BD Falcon™ disposable centrifuge tubes, polypropylene, conical bottom (BD Biosciences, Franklin Lakes, NJ, Cat. No. 352075)
1.0 mm glass beads (BioSpec Products, Inc., Bartlesville, OK, Cat. No. 11079110)
Milk powder
RNAse A (Qiagen, Cat. No. 19101)
Beckman centrifuge (Beckman Coulter, Fullerton, CA, Rotor No. SW28, Serial No. 880)
Sorvall centrifuge (Thermo Scientific, Rockford, IL, Cat. No. 46900)
Dri-Block® heaters (Techne, Ltd., Burlington, NJ, Cat. No. 6013502)
VWR® Vortex Mixer (VWR, West Chester, PA, Cat. No. 12620-838)
DU® 800 UV/Vis spectrophotometer (Beckman Coulter, Fullerton, CA, Cat. No. BK 512800)

Sonicator (Fisher Scientific, San Jose, CA, Model 100)
FPLC chromatographic system (GE Healthcare, Piscataway, NJ, Model AKTA)

Reagent Preparation

IPTG (1 M):
  238 mg/mL in H_2O, sterile filter, store in aliquots at –20°C
Lysis buffer (1 L):
  50 mM NaH_2PO_4 (6.90 g $NaH_2PO_4 \cdot H_2O$)
  300 mM NaCl (17.54 g NaCl)
  10 mM imidazole (0.68 g imidazole)
  Adjust pH to 8.0 using 10 M NaOH

Step-by-Step Protocol

Bacterial Cell Culture for Recombinant Protein Expression

Day 1 Streak plate of cells with vector for protein of interest (Note 2 in the Troubleshooting Guide) on LB-plate with appropriate antibiotic selection, and incubate the plate overnight at 37°C.

Day 2

1. Select five to six single colonies.
2. Inoculate two to five colonies into 2 mL cultures (2XYT + selective antibiotic).
3. Incubate the inoculated culture overnight at 37°C on a rotating shaker.

Day 3 (Mini-induction protocol to choose the most robust protein expression clone)

1. Take 100 μL from the saturated culture, and inoculate into 1 mL of fresh media (2XYT + selective antibiotic). Preserve the saturated cultures from Day 2, and store at 4°C.
2. Incubate for 1 h.
3. Add 1 mM IPTG to induce the expression of protein.
4. Incubate for 1–2 h.
5. Centrifuge the cells at 14,000 rpm in a 1.5 mL Eppendorf tube for 4 min.
6. Discard the supernatant.
7. Add 40 μL water, 4 μL of 4× LDS buffer, and 1 μL of 10× reducing agent to prepare the protein sample to be loaded onto an SDS gel, along with ten 1.0 mm glass beads to lyse the cells.
8. Vortex continuously at high speed for 1 min.
9. Place the sample in a heat block at 80°C for 10 min.
10. Vortex for another 1 min.
11. Briefly centrifuge tubes for 10 s.
12. Take the less viscous surface of supernatant, and add about 15 μL of sample in a gel lane.

13. Run SDS Page 4–12% Bis-Tris gel in MOPS running buffer.
14. Use Western blot to determine which clones expressed the greatest amount of protein.
15. Detect protein using anti-GST primary antibody (1:5000 in 5% milk in 1× TBS) and secondary antibody (1:2000 in 5% milk in 1 × TBS).
16. The clone with the highest expression level should be used for the 4 L preparation the following day.
17. Add 100–200 μL of this clone from Day 2 to each of four 100 mL cultures (2XYT + 50 μL/mL Ampicillin) in 250 mL culture flasks.
18. Incubate at 37°C overnight.

Day 4 (4 L protein preparation)

1. Transfer each of the now saturated 100 mL cultures into 1 L of fresh media (2YT + 50 μL/mL Ampicillin).
2. Incubate at 37°C and after 1 hour periodically check OD until OD reaches about 0.7–0.8 (see Note 3 in the Troubleshooting Guide).
3. Induce each flask with 10 mL of 100 mM IPTG to get a final concentration of 1 mM IPTG.
4. Incubate cultures for another 1–2 h.
5. Centrifuge cultures at 4000 rpm for 20 min at 4°C in 225 mL BD Falcon disposable centrifuge tubes.
6. Discard supernatants (Note 4).
7. Resuspend cells in lysis buffer at 2–5 mL per gram wet weight.
8. Add lysozyme to 1 mg/mL, and incubate on ice for 30 min.
9. Sonicate on ice (Setting #2, output at 50%, 200 W sonicator), six 10 s bursts with a 10 s cooling period in between.
10. Add RNAse A in a 1:1000 dilution; incubate on ice for 10–15 min.
11. Centrifuge in Sorvall centrifuge at 14,000 rpm for 5 min.
12. Save the supernatant.
13. Run supernatant through high-speed centrifuge, SW28 swinging bucket rotor, at 23,000 rpm for 30 min.
14. Protein is ready to be loaded on an FPLC column.
15. Allow column to run overnight, and test fractions using gel electrophoresis the following day by Western blotting for expression and quantitation of the target protein.

Protocol 6.3: Printing of Protein Microarrays

How It Works

Once target proteins are purified, they have to be tested to ensure that they are expressed as full-length molecules of predicted size that incorporate the epitope tag of interest. The different tags could be GST, V5, 6-histidine, or HRP. As per this protocol, proteins are tagged with GST. Any number of slides can be printed from 1–100 per

Alexa Ab	Alexa Ab	Buffer	Buffer	Printing Buffer	Printing Buffer	BLANK	BLANK	Ptn 1	Ptn 1
Human IgG 1	Human IgG 1	Human IgG 2	Human IgG 2	Human IgG 3	Human IgG 3	Ptn 2	Ptn 2	Ptn 3	Ptn 3
GST Tag	GST Tag	Ptn 4	Ptn 4	Buffer	Buffer	Ptn 5	Ptn 5	Ptn 6	Ptn 6
Ptn 7	Ptn 7	Ptn 8	Ptn 8	BLANK	BLANK	BLANK	BLANK	Buffer	Buffer
Ptn 9	Ptn 9	Printing Buffer	Printing Buffer	Alexa Ab	Alexa Ab	Buffer	Buffer	GST Tag	GST Tag
Ptn10	Ptn 10	Ptn 11	Ptn 11	Ptn 12	Ptn 12	Printing Buffer	Printing Buffer	Ptn 13	Ptn 13
Ptn14	Ptn 14	Ptn 15	Ptn 15	Ptn 16	Ptn 16	BLANK	BLANK	Ptn 17	Ptn 17
Ptn18	Ptn 18	PBS	PBS	Ptn 19	Ptn 19	Printing Buffer	Printing Buffer	Ptn 20	Ptn 20
BLANK	BLANK	Printing Buffer	Printing Buffer	Ptn 21	Ptn 21	BLANK	BLANK	Ptn 22	Ptn 22
Anti Human IgG 1	Anti Human IgG 1	Anti Human IgG 2	Anti Human IgG 2	Anti Human IgG 3	Anti Human IgG 3	Anti Human IgG 4	Anti Human IgG 4	Alexa Ab	Alexa Ab

Figure 6.2
A map of a subarray for custom proteins. A typical map for proteins to be printed on a subarray is represented in this figure. Proteins are printed in duplicates, and if area permits, then they can be printed in two different locations. A header for the subarray (Alexa-Ab for this array) is always printed in the same position for the correct orientation of the array. Positive controls such as human IgG and anti-human IgG are printed in three and four different concentrations, respectively. Negative controls such as Buffer (PBS) and printing buffer for custom proteins are printed at various locations on the array for detection of uneven scanning.

run of the instrument. Initially, a few slides are printed and processed until the slides are optimized for the correct size spot and the distance between spots (see Note 5 in the Troubleshooting Guide). For one run, 80 slides is an optimal number. The correct humidity (60%) should be maintained during each print run. Temperature should be kept constant at 6°C for all print runs. This ensures that all the slides are subject to the same conditions. The slides are held in place by vacuum on the printing stage and are numbered to reflect their print batch (date) and position in the printing sequence. Once the proteins are printed, the first slide, one slide midway, and the last slide of the lot should be analyzed for correct printing of the array, by processing them with an antibody against the tag on the protein. Because GST is tagged to the proteins, our initial quality-check processing is done using GST antibodies.

A map of the proteins, showing their exact position on the subarray, should be prepared before printing (Figure 6.2). Each protein should be printed in duplicate and preferably also on two different locations of the subarray. The secondary antibody fluorochrome should be used as a header for all the subarrays and preferably printed in the same position in each subarray (Note 6). Plain phosphate-buffered saline (PBS) can be used as a negative control, and spots with no protein printed are also included in the subarray. Imidazole may be used in our laboratory as a component of the print buffer, to enable visualization of the printed proteins at the respective wavelength even without processing the slide. The spot diameter used is 400 μm in size (pinhead size) with 360 μm spacing between each printed spot. (The smaller the spot, the more spots one can print in a subarray, making it a high-density array; however, care should be taken to avoid merging of spots.) Solid or quill pins can be used for printing. We use solid pins for printing the H-Y array with four pins loaded at a time. Between each print for different proteins, the pins are washed five times with cold 1× PBS (Note 7). Once the proteins are printed, allow them to dry for 1 h before storing them at –20°C.

Required Materials

Precoated glass slides with nitrocellulose FAST Slide-1 Pad (Whatman, Inc., Florham Park, NJ, Cat. No. 10484182)
Proteins to be printed (1 mg/mL)
Contact protein printer (Bio-Rad, Hercules, CA, Model ChipWriter Pro)
Stealth Microarray printhead for 32 pins (Telechem International, Sunnyvale, CA, Cat. No. SPH32)
Stealth Micro spotting prints (Telechem International, Sunnyvale, CA, Cat. No. SMP9)
384-well amplification plates (Nunc, VWR, West Chester, PA, Cat. No. 230582)
PBS (1×)

Reagent Preparation

Print buffer
1× PBS
Imidazole (Sigma-Aldrich, Cat. No. I5513)
Proteins were dissolved in 1× PBS with 100 mM imidazole before printing to get a final concentration of 1 mg/mL.

Step-by-Step Protocol

1. Place 10 μL of 1 mg/mL protein in five rows of a 384-well plate with a distance of two rows between each row of the plate filled with the protein solution.
2. Adjust slides in a row on the contact protein printer.
3. Set the printer to load a precalculated amount of proteins (2 μL) from the 384-well plate onto the glass slide.

4. Make sure the orientation of the slide coincides with the first protein to be printed.
5. After printing, place each slide in a slide holder box, seal it, and store at −80°C (see Note 8 in the Troubleshooting Guide).

Protocol 6.4: Assay Development

How It Works

Protein microarrays are probed with test samples and control samples. If there are antibodies against any of the printed proteins, then an antigen–antibody complex should form. This is detected using a secondary antibody conjugated to a fluorochrome (Figure 6.2). The fluorescence intensity is detected using a microarray scanner. The values obtained are normalized for each protein microarray against the background and also with positive and negative controls. The normalized readings are compared for differences between the test samples and control samples. This protocol can be used for commercial slides as well as for custom-made protein microarrays.

Required Materials

Anti-human antibody–AlexaFluor conjugate (Molecular Probes, Cat. No. A21445)
10× PBS, pH 7.4 (GIBCO, Invitrogen, Carlsbad, CA, Cat. No. 70011-044)
1× PBS solution, pH 7.4
Tween 20 (American Bioanalytical, Natick, MA, Cat. No. AB02038)
L-glutathione reduced (Sigma, Cat. No. G-4251)
BSA, protease free, 30% solution (Sigma, Cat. No. A-8577)
1% BSA solution
4-well trays (quadriPERM 4-chamber culture dish) (Greiner ISC BioExpress, Kaysville, KY, Cat. No. 96077307)
GenePix 4000B microarray scanner (Molecular Devices Corporation, Sunnyvale, CA)
Eppendorf centrifuge (5810) (Fisher Scientific, San Jose, CA, Cat. No. 05-400-60)
Lab rotator (Lab-Line Instruments, Melrose Park, IL, Cat. No. 1314)
GenePix Pro software (Molecular Devices Corporation, Sunnyvale, CA)

Reagent Preparation

Blocking buffer: Prepare 1 L containing 50 mM HEPES, pH 7.5; 200 mM NaCl; 0.08% Triton X-100; 25% glycerol; 20 mM reduced glutathione; and 1.0 mM DTT. Add 1% bovine serum albumin (BSA) and 1 mM DTT to solution just prior to use. Adjust the pH to 7.5–8 with 10 M NaOH (see Note 9 in the Troubleshooting Guide).

PBST buffer: Prepare 1 L containing 1× PBS, 1% BSA, and 0.1% Tween 20. Add 1% BSA to solution just prior to use.

AlexaFluor conjugate solution: Dilute anti-human antibody–Alexa Fluor® 647 conjugate (2 mg/mL) to 1.0 μg/mL in PBST buffer.

Step-by-Step Protocol

Probing Protein Microarrays

1. Experiment should be carried out at 4°C.
2. Place the blocking solution (5 mL) in a 4-well tray.
3. Immerse a protein microarray in the blocking solution with the side of proteins printed facing up (see Note 10 in the Troubleshooting Guide).
4. Gentle agitation should be applied to the tray for 1 h.
5. Remove the blocking buffer by aspiration.
6. Dilute serum/plasma at 1:150 (v/v) in PBST buffer.
7. Add the diluted serum/plasma to the tray containing the protein microarray.
8. Incubate for 90 min at 4°C with gentle agitation.
9. Remove the diluted serum sample by aspiration.
10. Wash the protein microarray with 5 mL of PBST buffer for 5 min with gentle agitation.
11. Remove the PBST buffer by aspiration.
12. Repeat steps 10 and 11 four more times (Note 11).
13. Add 5 mL of diluted secondary antibody diluted in PBST buffer as described earlier.
14. Incubate the protein microarray at 4°C for 90 min with gentle agitation.
15. Wash the protein microarray with 5 mL of PBST buffer for 5 min with gentle agitation.
16. Remove the PBST buffer by aspiration.
17. Repeat steps 15 and 16 four more times.
18. Place protein microarray slides in slide holders and centrifuge at 500 rpm for 1 min at room temperature.
19. Store slides in a slide container box in the dark at 4°C until read. Slides should not be exposed to light. They should be read within 24 h after completing the probing of the protein microarray.

Scanning the Arrays Using a Microarray Scanner

20. Acquire the data and analyze them using any software package for microarray analysis.
21. For GenePix 4000B microarray scanner, start the software for acquisition of the data for the protein microarray.
22. The software on the PC should be attached to the microarray reader.
23. Scan the slides using, preferably, GenePix 4000B microarray scanner.
24. Place the slide in the microarray holder on the instrument.
25. The nitrocellulose slide surface coated with proteins should face the laser source.
26. Once the microarray is in place, choose the following setting:
    a. Wavelength (if using Alexa 647): 647 nm, or choose the appropriate wavelength (Note 12 in the Troubleshooting Guide) based on the fluorochrome selected
    b. Photo multiplier tube (PMT) gain: 600

c. Laser power: 100%
d. Pixel size: 10 μm
e. Lines to average: 1.0
f. Focus position: 0 μm

27. A preview scan of the protein microarray should be performed to adjust the PMT.
28. Typically, the highest concentration of IgG should be at saturation (60,000 fluorescence units), and the subsequent highest concentration should be in the range of 48,000 to 50,000 (Figure 6.3; Note 13).
29. A .gal file should be created with the software package. This file would contain the identity of the proteins in each location.
30. Open the correct .gal file, and scan the protein microarray slide at the selected PMT gains.
31. A .tif file for the high-resolution image will be obtained along with a .txt file containing the fluorescent readings of each spot with its location and identification of the proteins in each spot.
32. Compare fluorescent values between protein microarrays of the test and control samples.

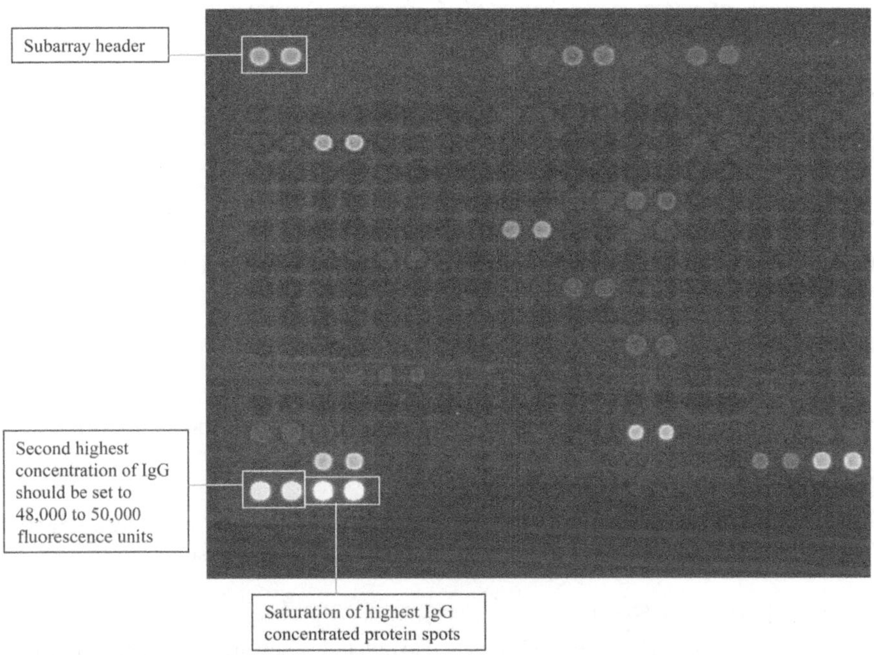

Figure 6.3
Adjustment for PMT during a prescan on one protein subarray of ProtArray™. A prescan before the final reading of a protein microarray is highly recommended. For this particular array, during the prescan, the PMT is dialed until the highest concentration of anti-human IgG printed reached 60,000 PMT units and the subsequent lower concentration of anti-human IgG is at 50,000 PMT units. During the prescan, the array header should fluoresce, indicating the start of each subarray.

Protocol 6.5: Storage

The processed slides can be stored for up to a year in the dark in an air-tight container at 4°C. The proteins identified as target proteins can be further validated by Western blotting or ELISA, thus confirming the findings in two different systems.

Troubleshooting Guide

Three major problems are encountered: high background, low signal intensities, or uneven background. In order to avoid these problems, the following points should be kept in mind:

Note 1: Smudges

Serum should be centrifuged as described earlier to avoid any impurities or floccules so as to avoid blotches on the surface of the slide.

Note 2: No Viable Cultures

Inoculate cultures with small scrapings using a sterile pipette tip of cells from the frozen stock of vector cells from −80°C freezer. Do not allow cells to thaw and refreeze; this will kill them.

Note 3: Low Protein Yield

Absorbance is assessed at 595 nm with blank as fresh 2XYT + 50 μL/mL Ampicillin media. Check the OD between 60 and 90 min.

Note 4: Degraded Protein

Work with cell pellets on ice or at 4°C at all times after this step to prevent protein degradation.

Note 5: Protein Spot Distribution

Always run a sample test after printing 1–2 slides to make sure the spot size is correct, with each spot remaining distinct and not merging with the neighboring spot.

Note 6: Misaligned Spots

Always print positive controls in the same place throughout the subarrays to help one identify and validate the .gal file results with the actual processed protein spot.

Note 7: Irregular Spots

The printer pins should be thoroughly cleaned to preclude uneven protein deposition in each spot. Some examples are protein spots that when processed appear doughnut shaped.

Note 8: Empty Areas

The slide surface with the proteins printed should not be touched from now on or during the experiment so as to avoid scraping off the printed proteins.

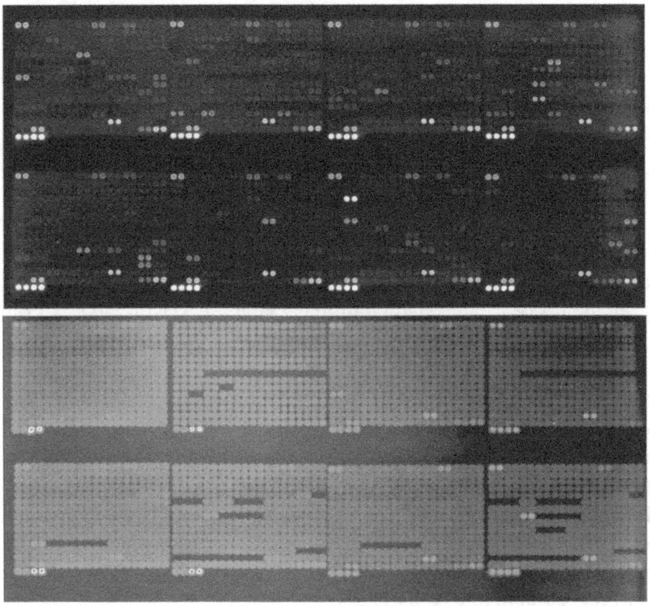

Figure 6.4
Comparison of slides. Care should be taken to prevent the slide from drying completely. The lower panel shows a slide that dried during processing, whereas the upper panel slide was processed correctly.

Note 9: Low Intensity Readings

The pH of all solutions should be optimal, and all buffers should be freshly prepared and preferably filtered with no precipitates in them so as to avoid low fluorescence readings on the scanner.

Note 10: Uneven Backgrounds

High or uneven backgrounds may arise if the slide is allowed to dry. The slide should not be allowed to dry during any stage of the experiment after Step 3 in Protocol 6.4 (Figure 6.4).

Note 11: High Backgrounds

Increase or decrease the number of washes depending on your results. The numbers of washes mentioned in Protocol 6.4 are optimal, but a particular serum sample may need higher or lower stringency.

Note 12: No Signal

Read the slides at the correct wavelength filters.

Note 13: High Signal-to-Noise Ratio

Dial up the PMT so that the positive controls are near saturation (but not saturated) to get a good signal-to-background ratio.

References

1. Ekins, R. P. (1989). Multi-analyte immunoassay. *Journal of Pharmaceutical and Biomedical Analysis*, 7: 155–168.
2. Duyk, G. M. (2002). Sharper tools and simpler methods. *Nature Genetics*, 32(Suppl): 465–468.
3. Margreth, A. et al. (1963). A morphological and biochemical study on the regulation of carbohydrate metabolism in the muscle cell. *Experimental Cell Research*, 32: 484–509.
4. Laemmli, U. K. (1970). Cleavage of structural proteins during the assembly of the head of bacteriophage T4. *Nature*, 227: 680–685.
5. Yamasaki, C. et al. (2008). The H-Invitational Database (H-InvDB), a comprehensive annotation resource for human genes and transcripts. *Nucleic Acids Research*, 36: D793-9.
6. Kung, L. A. and Snyder, M. (2006). Proteome chips for whole-organism assays. *Nature Reviews Molecular Cell Biology*, 7: 617–622.
7. Lynch, M. et al. (2004). Functional protein nanoarrays for biomarker profiling. *Proteomics*, 4: 1695–1702.
8. Templin, M. F. et al. (2003). Protein microarrays: promising tools for proteomic research. *Proteomics*, 3: 2155–2166.
9. Grubb, R. L. et al. (2003). Signal pathway profiling of prostate cancer using reverse phase protein arrays. *Proteomics*, 3: 2142–2146.
10. Kusnezow, W. et al. (2005). Solid supports for protein microarrays and related devices. In *Protein Microarrays*. M. Schena, Ed. (pp. 247–283), Sudbury, MA: Jones and Bartlett.
11. Gutmann, O. R. et al. (2005). Fast and reliable protein microarray production by a new drop-in-drop technique. *Lab on a Chip*, 5: 675–681.
12. Haab, B. B. et al. (2001). Protein microarrays for highly parallel detection and quantitation of specific proteins and antibodies in complex solutions. *Genome Biology*, 2: RESEARCH0004.
13. Goulmy, E. (1996). Human minor histocompatibility antigens. *Current Opinion in Immunology*, 8: 75–81.
14. Miklos, D. B. et al. (2004). Antibody response to DBY minor histocompatibility antigen is induced after allogeneic stem cell transplantation and in healthy female donors. *Blood*, 103: 353–359.
15. Miklos, D. B. et al. (2005). Antibody responses to H-Y minor histocompatibility antigens correlate with chronic graft versus host disease and disease remission. *Blood*, 105: 2973–2978.

Chapter 7

Bead Arrays

An Introduction to Multiplexed Bead-Based Assays for Proteins

Yong Song

Contents

Introduction

Multiplex, a term once only familiar to moviegoers from the film *Multiplicity* (1996), has become a commonly used word in the biomedical testing field. After more than a decade of development, multiplexed assay technology has become a regular laboratory tool in many research areas, including immunology, oncology, molecular cell biology, toxicology, neurobiology, etc. A multiplexed assay measures multiple analytes simultaneously in a single reaction. The wealth of information provided by these kinds of high-content assays has revolutionized the way in which drug discovery, drug development, vaccine testing, and diagnostics are conducted. One of the biggest challenges in today's drug development arena, which includes small molecule, biologics, and vaccine development, is the comprehensive analysis of drug effects. Such effects are both *in vitro* and *in vivo*, intercellularly or intracellularly in nature. Analyzing one analyte at a time has become a bottleneck in understanding the mechanism of action, efficacy, and side effects related to these therapeutic candidates.

Multiplexed assay technology developed for immunoassay to date can be categorized into two basic groups. One is bead-based arrays with antibodies conjugated on the beads and analyzed by flow cytometry; the other is spot-based microarrays in which antibodies are printed onto a planar solid support and analyzed by imaging. The bead-based assays are largely represented by two major platforms: Luminex xMAP and BD Biosciences' cytometric bead assays. Other bead-based assay providers include Beckman Coulter's Bender Flowcytomix assays and Bangs Laboratories' QuantumPlex beads. There are also commercial spot-based assays such as those offered by MesoScale Discovery and Thermo Scientific's SearchLight. The Luminex xMAP technology requires a specific cytometer, Luminex reader, for bead population identification and signal detection.

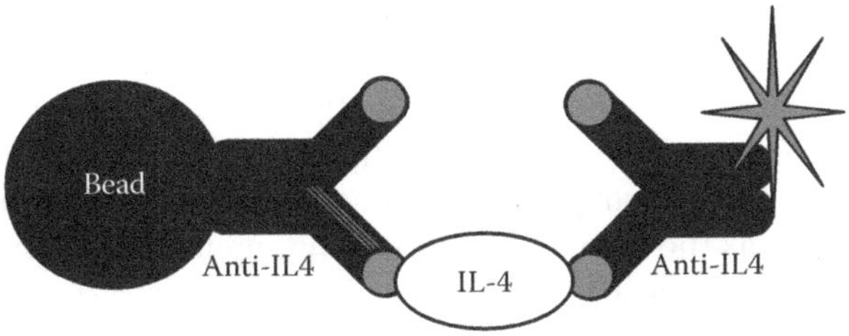

Figure 7.1
Bead-based immunoassay complex.

Assays such as BD's CBA and Beckman Coulter's Flowcytomix are designed for use on a variety of flow cytometer models that are available from different manufacturers.

In this chapter, we focus only on multiplex assay technologies for immunoassays based on platforms that utilize beads with color differentiation detected by flow cytometry. Captured molecules on the beads are detected by fluorescence labeling, and the signal is analyzed by flow cytometry. Bead-based multiplex platforms have the capacity to measure up to 100 analytes simultaneously. One of the platforms for bead- based multiplex assays is Luminex's xMAP technology developed by Luminex Co. (Austin, TX). This technology utilizes distinct fluorophore-coded microspheres with discrete red and near-infrared fluorescent-dye ratios that function as zip codes or addresses for individual analytes. Each individual bead set is conjugated with a capture reagent specific to a particular bioassay. A number of distinct bead sets with specific reagents can then be mixed in a single reaction tube or well of a multiwell plate to achieve multiplex assays. The Luminex xMAP analyzer consists of two lasers as excitation sources. One laser excites the beads' internal dyes, which identify each bead set, whereas the other laser excites a reporter dye such as R-phycoerythrin captured during the immunoassay. A typical sandwich immunoassay for a specific analyte is illustrated in Figure 7.1.

Flow cytometer manufacturers such as Beckman Coulter and BD Biosciences also provide bead-based assays that are based on bead size and fluorescent intensity to differentiate bead populations. The reaction principle is basically the same as that of the Luminex technology. The key advantage is that the assays can be run on many different flow cytometry platforms.

In comparison with the planar surface, spot-based arrays, bead assays offer faster reaction kinetics. This is because microspheres are of smaller size and thereby provide a large surface area. As the reaction is solution based, with all reaction elements moving constantly during the reaction period, there is higher frequency for collision. This leads to more efficient capture and improved binding kinetics. Compared to conventional ELISAs, microsphere-based multiplexed immunoassays usually have better assay sensitivities and are as reproducible and reliable as ELISAs. Furthermore, multiplexed assays have the clear advantage of detecting multiple analytes simultaneously in a small sample volume, which requires considerably less time and reduced reagent consumption. This makes bead-based assays especially beneficial in pediatric practices and *in vivo* mouse studies.

Similar to conventional ELISAs, quantitative analysis of the microsphere-based immunoassays requires a robust curve-fitting program for analysis (Gottschalk and Dunn, 2005). Beckman Coulter supplies an analysis software package, FlowCytomix Pro, for their bead-based multiplexed assays, which can perform five-parameter logistic analyses and has data-reduction and report-building capabilities. Other vendors such as Bio-Rad or MariaBio, who supply the Luminex xMAP readers, also provide curve fitting and data analysis in their systems. Both Bio-Rad and MariaBio further provide 21CFR Part 11 compliance software to support drug development.

Many different assays are applicable on bead-based systems. Among others, these include cytokine; growth factors and other secreted proteins; cancer and cardiac markers; intracellular protein activation (cell signaling); specific IgG and IgE detections for autoimmune, allergy, and autoantibody determinations; transcription factor activation; mircoRNA; SNP; gene expression; and genotyping.

In this report, we discuss the assay development protocol of the bead-based sandwich immunoassay (Figure 7.1). It requires two specific antibodies against two different epitopes of a protein of interest (see Notes 1 and 2 of the Troubleshooting Guide regarding antibody selection).

List of Protocols

Protocol Number	Name
7.1	Bead Selection
7.2	Buffer Exchange
7.3	Protein Concentration Determination
7.4	Bead Conjugation
7.5	Antibody Biotinylation
7.6	Sandwich Immunoassays

Protocol 7.1: Bead Selection

How It Works

The most popular choice of beads for immunoassays is carboxylated group-coated beads. The conjugation process is straightforward and does not take too long. It is fairly easy to scale up for production purpose. There are also a few other bead selections to consider. For example, serological studies can use beads that are designed for this purpose, such as Luminex's SeroMAP microspheres, which are specially blocked to reduce background signal when testing plasma or serum samples. Luminex and Bangs Laboratories provide a variety of magnetic beads. Assays with magnetic beads do not require the use of filter plate technology for bead washing. Such assays are easily adapted for use with automated liquid-handling robots.

Protocol 7.2: Buffer Exchange

How It Works

This procedure is needed if antibodies provided are stored in a buffer solution that contains Tris or other buffer components that contain primary amines. These will interfere with the protein conjugation to the carboxyl group coated microspheres.

Required Materials

Equipment: Microfuge 18 centrifuge (Beckman Coulter, Fullerton, CA) or equivalent.
Material: Zeba™ Desalt Spin Columns, 0.5 mL (Pierce, Rockford, IL; Part No. 89882).
Note: Micro Bio-Spin P-6 Tris chromatography columns (Bio-Rad, Hercules, CA) can also be used for this purpose.
1.5–2.0 mL microcentrifuge collection tubes.
Phosphate-buffered saline (PBS) 7.4 (Invitrogen, Carlsbad, CA, Part No. 10010-023) or equivalent.

Step-by-Step Protocol

The following procedure is modified from the protocol provided in the product insert of the Zeba Desalt Spin Columns.

Protein Desalting Spin Column Preparation

1. Remove column's bottom closure, and loosen cap (do not remove cap).
2. Place column in a 1.5–2.0 mL microcentrifuge collection tube.
3. Centrifuge at 1500× g for 1 min to remove storage solution.
4. Place a mark on the side of the column where the compacted resin is slanted upward. Place column in the microcentrifuge with the mark facing outward in all subsequent centrifugation steps.
5. Add 300 μL of PBS on top of the resin bed. Centrifuge at 1500× g for 1 min to remove buffer.
6. Repeat step 5 two to three additional times, discarding buffer from the collection tube.

Sample Loading

7. Place column in a new collection tube, remove cap, and apply 30–130 μL of sample to the top of the compact resin bed.
8. (Optional) For sample volumes <70 μL, apply a 15 μL stacker of PBS to the top of the gel bed after the sample has fully absorbed, to ensure maximal protein recovery.
9. Centrifuge at 1500× g for 2 min to collect the sample.
10. Discard desalting column after use.

Protocol 7.3: Determination of Antibody IgG Concentration

How It Works

In order to find the optimal protein-to-bead ratio necessary to achieve maximal conjugation efficiency, protein concentrations of the antibodies should be determined by using either the Lowry or Bradford protein assay method. Bovine gamma globulin should be used as the standard. The Bradford protein assay method is a simple and quick procedure that works well with antibodies following buffer exchange.

Required Materials

Equipment:

- Barnstead/Lab-Line Titer Plate Shaker (Thermo Scientific, Waltham, MA) or equivalent.
- PARADIGM plate reader with Absorbance Detection Cartridge (Beckman Coulter, Fullerton, CA) or equivalent.

Materials:

- Coomassie (Bradford) protein assay kit (Pierce, Rockford, IL, Part No: 23200).
- Bovine Gamma Globulin Fraction II (BGG) Pre-Diluted Protein Assay Standards (Pierce, Rockford, IL, Part No. 23213) in the range of 125–2000 μg/mL.
- Phosphate-buffered saline (PBS) 7.4 (Invitrogen, Carlsbad, CA, Part No. 10010-023) or equivalent. It should be the same buffer used for the buffer exchange step.

Step-by-Step Protocol

The following standard microplate protocol with a working range from 100–1500 μg/mL is modified from the protocol provided in the product insert of the Coomassie (Bradford) protein assay kit.

1. Pipette 5 μL of each standard or unknown sample, including a blank buffer control, into appropriate microplate wells.
2. Add 250 μL of the Coomassie reagent to each well using a calibrated multichannel pipette, and mix on a plate shaker for 30 min at room temperature.
3. Remove the plate from the shaker.
4. Measure the absorbance at or near 595 nm with a plate reader.
5. Subtract the average 595 nm measurements for the blank replicates from the 595 nm measurements of all other individual standard and unknown sample replicates.
6. Prepare a standard curve by plotting the average blank-corrected 595 nm measurement for each BSA standard versus its concentration in μg/mL with a four-parameter (quadratic) or best-fit curve-fitting algorithm. Use the standard curve to determine the protein concentration of each unknown sample.

Protocol 7.4: Bead Conjugation

How It Works

The procedure is modified from the Luminex user manual and a procedure described by Skogstrand et al. (2005). For each assay, it may be necessary to optimize the protein-to-bead conjugation ratio to better meet the analyte's assay range.

Required Materials

Equipment:
- Vortex mixer (VWR, West Chester, PA) or equivalent
- Barnstead/Thermolyne Labquake Rotisserie Rotator (Thermo Scientific, Waltham, MA) or equivalent
- Hemocytometer or particle counters such as Z counters (Beckman Coulter, Fullerton, CA)
- Microfuge 18 centrifuge (Beckman Coulter, Fullerton, CA) or equivalent
- Bransonic ultrasonic cleaners (Branson, Danbury, CT) or equivalent

Materials:
- Microcentrifuge tube (USA Scientific, Ocala, FL, Part No. 1415-2500)
- Carboxylated polystyrene microspheres (Luminex Corporation, Austin, TX)
- Antibody for coupling
- Monobasic sodium phosphate
- *N*-hydroxysulfosuccinimide (NHSS) MW 217.3
- 1-ethyl-3-(3-dimethylaminopropyl)carbodiimide hydrochloride (EDC)
- 10× phosphate-buffered saline (PBS) pH 7.4 (Bio-Rad, Hercules, CA, Cat. No. 161-0780)
- Tween 20
- Bovine serum albumin (BSA)
- Sodium azide
- 1 N NaOH
- 1 N HCl

Reagent Preparation

Activation buffer: Prepare 0.1 M sodium phosphate, pH 6.0, by dissolving 12 g monobasic sodium phosphate in 800 mL water. Adjust pH to 6.0 with 1 N HCl, and then bring final volume to 1 L. Check final pH. Filter through a 0.2 μm filter. Store up to 6 months at 4°C.

NHSS solution: Dissolve 50 mg in 1 mL activation buffer. Mix well. NHSS storage container must be brought to room temperature prior to opening. The reagent should be prepared immediately before use.

EDC solution: Dissolve 50 mg EDC in 1 mL activation buffer. Mix well. EDC storage container must be brought to room temperature prior to opening. The reagent should be prepared immediately before use.

Wash buffer: Prepare PBS, pH 7.4, 0.05% Tween 20, by mixing 100 mL 10× PBS with 0.5 g Tween 20. Add 800 mL water. Adjust pH to 7.4 with 1 N HCl or 1 N NaOH. Bring final volume to 1 L. Check final pH. Filter through a 0.2 μm filter. Store up to 6 months at room temperature.

Blocking/storage buffer: Prepare PBS, pH 7.4, 10 mg/mL BSA, 0.02% sodium azide, by dissolving 10 g BSA in 100 mL 10× PBS. Add 0.2 g sodium azide in 800 mL water. Adjust pH to 7.4 with 1 N HCl or 1 N NaOH. Bring final volume to 1 L. Check final pH. Filter through a 0.2 μm filter. Store up to 6 months at 4°C.

Step-by-Step Protocol

1. Calculate the amount of antibody and beads needed.
   The authors found that for the 5.6 μm carboxylated microspheres, 3–5 μg IgG per 1 million beads works well for most of the cytokine assays.
2. Vortex the bead stock for 10–15 s.
3. Further resuspend the beads with sonication for 20–30 s.
4. Vortex for another 10–15 s.
5. Determine the bead concentration by using a hemocytometer or Z counter according to the manufacturer's instruction manual.
6. Transfer appropriate amount of beads from the homogeneous bead stock into a microcentrifuge tube (reaction tube).
7. Centrifuge the reaction tube for 1 min at 10,000× g.
8. Aspirate the supernatant using a pipette. Avoid disturbing the bead pellet while pipetting.
9. Wash twice with 100 μL the activation buffer.
10. Resuspend the beads in 80 μL of activation buffer with sonication for 20–30 s.
11. Vortex for 20–30 s.
12. Prepare a 50 mg/mL NHSS solution immediately before use.
13. Add 10 μL of the 50 mg/mL NHSS solution to the reaction tube.
14. Vortex for 10 s.
15. Prepare a 50 mg/mL EDC solution immediately before use.
16. Add 10 μL of the 50 mg/mL NHSS solution to the reaction tube.
17. Vortex for 10 s.
18. Incubate for 20 min in the dark at room temperature.
19. Centrifuge the activated beads for 1 min at 10,000× g.
20. Aspirate the supernatant.
21. Wash the beads twice with 300 μL 1× PBS.
22. Aspirate the supernatant after the last wash.
23. Add 400 μL of PBS to the reaction tube.
24. Vortex for 30 s.
25. Add appropriate amount of stock antibody diluted to a total of 100 μL in PBS.
26. Mix gently.
27. Rotate the mixture for 2 h in the dark at room temperature.
28. Centrifuge the microcentrifuge tube with beads and antibody for 2 min at 10,000× g.
29. Aspirate all supernatant.

30. Wash the beads twice with 500 μL of the wash buffer.
31. Suspend the conjugated beads in 200 μL of blocking/storage buffer.
32. Determine the bead concentration by using a hemocytometer or Z counter according to the manufacturer's instruction manual.
33. Adjust the bead concentration to 5,000,000 beads/mL (100×) for easy assay setup.
34. Label and store the conjugated beads at 4°C. Protect from light.

Protocol 7.5: Biotinylation of Antibodies

How It Works

Biotinylation of IgG can be achieved by using sulfo-NHS-LC-Biotin. The NHS ester reacts with primary amines at the amino terminus and those of the epsilon amine of lysine residues present in immunoglobulin G. It has been suggested that a molar ratio of biotin to protein of 22:1 yields biotin substitution levels of 8–14 mol of biotin per mole of IgG. The authors found that a 40:1 molar ratio of sulfo-NHS-LC-Biotin to IgG gives a biotin-substituted level of 10 mol biotin per mole IgG when performing a 1–3 mg IgG scale of biotinylation. It is strongly recommended that biotin-substituted levels should be optimized by varying molar ratio of biotin to IgG with each antibody, especially for large-scale (more than 10 mg IgG) biotinylations. The No-Weigh™ package of this biotinylation reagent is recommended for consistent biotinylation yield and efficiency because NHS esters are hydrolyzed very quickly.

Required Materials

Equipment:
- Vortex mixer (VWR, West Chester, PA) or equivalent
- Microfuge 18 centrifuge (Beckman Coulter, Fullerton, CA) or equivalent

Materials:
- No-Weigh Sulfo-NHS-LC-Biotin (Pierce, Rockford, IL, Part No. 21327)
- Zeba Desalt Spin Columns, 0.5 mL (Pierce, Rockford, IL; Part No. 89882)
- Phosphate-buffered saline (PBS) 7.4 (Invitrogen, Carlsbad, CA, Part No. 10010-023) or equivalent

Reagent Preparation

Antibody: If it is dissolved in Tris or other amine-containing buffers, buffer exchange to PBS should be conducted (see Protocol 7.2). If protein concentration is too low (<1.5 mg/mL), a concentration step using, for example, a Centricon YM30 (Millipore, Billerica, MA; Part No. 4242AM) to concentrate the protein concentration to 2 mg/mL is recommended.

No-Weigh Sulfo-NHS-LC-Biotin: Prior to opening, the vial must be brought to room temperature.

Calculations (40:1 molar ratio of sulfo-NHS-LC-Biotin to IgG): For every 1000 μg mAb in a reaction volume of 500 μL, add 27 μL of 10 mM Sulfo-NHS-LC-Biotin.

Step-by-Step Protocol

1. Add 180 μL of ultrapure water to one of the 1 mg No-Weigh Sulfo-NHS-LC-Biotin microtubes immediately before use.
2. Add the appropriate volume of 10 mM biotin reagent solution to the protein solution.
3. Incubate reaction on ice for 2 h or at room temperature for 30 min.
4. Purify the biotinylated IgG using the same procedure described in Protocol 7.3.
5. Determine protein concentration using the Bradford protein assay method.
6. Adjust the protein concentration to 0.5 mg/mL with PBS.
7. Optional: Determine the level of biotin incorporation in IgG using Biotinylation EZ™ Biotin Quantitation Kit (HABA assay; Pierce, Rockford, IL, Part No. 28005).

Protocol 7.6: Sandwich Immunoassays

How It Works

As illustrated in Figure 7.1, similar to the principle of a sandwich ELISA, beads conjugated with a specific antibody serve as capture molecules to trap the protein of interest, such as a cytokine, in the sample. The amount of the analyte captured on the bead surface is detected via a biotinylated antibody against a secondary epitope of the protein, followed by a streptavidin-R-phycoerythrin treatment. The fluorescent intensity of R-phycoerythrin on the beads is quantified on the Luminex reader or a flow cytometer. Concentrations of the protein of interest in the samples can be obtained by comparing the fluorescent signals to those of a calibration curve generated from a serial dilution of a known concentration of the analyte.

Required Materials

Equipment:

- Barnstead/Lab-Line Titer Plate Shaker (Thermo Scientific, Waltham, MA) or equivalent
- Multiscreen vacuum manifold (Millipore, Billerica, MA) with vacuum source
- Bio-Plex (Bio-Rad, Hercules, CA) or Luminex xMAP (Luminex Corp, Austin, TX) or equivalent for Luminex bead-based assays
- FC500 (Beckman Coulter, Fullerton, CA), Quanta SC (Beckman Coulter, Fullerton, CA), or FACSArray (BD Biosciences, San Jose, CA) or other flow cytometers for BD CBA Flex, Bender Flowcytomix, or Bangs QuantumPlex bead-based assays

Materials:

- Samples: Cell culture supernatant, serum, plasma, etc.
- Recombinant or purified antigens as standards
- Antibody-coupled microspheres for each antigen
- Affinity-purified biotinylated detection antibodies specific for each antigen
- Multiscreen microtiter filter plates (MABVN 1.2 μm; Millipore, Billerica, MA)
- Aluminum foil lids (Beckman Coulter, Fullerton, CA, Part No. 538619)
- Streptavidin-R-phycoerythrin (Prozyme, San Leandro, CA) or equivalent

Reagent Preparation

Wash buffer: Prepare PBS, pH 7.4, 0.05% Tween 20, by mixing 100 mL 10× PBS with 0.5 g Tween 20. Add 800 mL water. Adjust pH to 7.4 with 1 N HCl or 1 N NaOH. Bring final volume to 1 L. Check final pH. Filter through a 0.2 μm filter. Store up to 6 months at room temperature.

Diluents: See the section "Troubleshooting Guide" for details regarding preparation of all diluents.

Protocol 7.6.1: Standard Preparation

1. Calculate the dilution volume of the highest concentration of the standard curve according to the stock antigen concentration. For example, if the stock concentration of IL-4 is 500,000 pg/mL and the desired highest concentration of the standard curve is 5,000 pg/mL, combine 5 μL of the stock IL-4 with 495 μL of the standard diluent.
2. Prepare the following serial dilutions (1:3 serial dilutions; 200 μL in total, enough for duplicated wells). Mix each addition by pipetting up and down 4–6 times. Use new pipette tips after each addition to avoid contamination from one concentration to the other (Table 7.1).

TABLE 7.1
Preparation of Antigen Standard Curve

Standards	Amount from Previous Standards (μL)	Standard Diluent (μL)
Standard 1 (Undiluted)		Prepared in the previous step
Standard 2 (1/3)	100	200
Standard 3 (1/9)	100	200
Standard 4 (1/27)	100	200
Standard 5 (1/81)	100	200
Standard 6 (1/243)	100	200
Standard 7 (1/729)	100	200
Standard 8 (blank)	0	200

Protocol 7.6.2: Antibody-Coupled Bead Working Suspension Preparation

1. Determine the number of wells needed for the experiment.
2. Calculate the amount of stock beads needed for each well:

   For each well, 2500 beads per each bead population is needed. The total volume for each well is 50 μL. Hence, the working concentration of each bead population is 50,000 beads/mL. The stock bead concentration after conjugation is 5,000,000 beads/mL, that is, 100× stock (see Protocol 7.4). Therefore, for each well, 0.5 μL of bead stock of each population is required.

3. Calculate the amount of wash buffer needed for each well for bead dilution.
   Volume of wash buffer in each well = 50 μL – 0.5 μL × number of analytes
   For example, if the multiplex assay panel consists of five analytes (IL-2, IL-3, IL-4, IL-10, and IL-12), the amount of wash buffer needed for each well:
   50 μL – 0.5 μL × 5 = 47.5 μL.
4. Vortex each bead stock vial until the beads are in suspension.
5. Add the required amount of beads of each population into the required amount of wash buffer.
   a. Required amount of beads = 0.5 μL × number of wells
   b. Required volume of wash buffer = result of step 3 × number of wells
   Note: When using a multichannel transfer pipette, 10 additional wells of reagents are required for pipetting loss and 5 wells if a single channel pipette is used.
6. Mix by vortexing.

Protocol 7.6.3: Biotinylated Detection Antibody Working Solution Preparation

1. Determine the number of wells needed for the experiment.
2. Calculate the amount of stock biotinylated antibody needed for each well:
   The working concentration of biotinylated antibody is 1 μg/mL. The total volume for each well is 50 μL. The stock biotinylated antibody concentration is 0.5 mg/mL, that is, 500× stock (see Protocol 7.5). Therefore, for each well, 0.1 μL of each biotinylated antibody stock is required.
3. Calculate the amount of biotinylated antibody diluent needed for each well for preparation of the working solution.
   Volume of biotinylated antibody diluent in each well = 50 μL – 0.1 μL × number of analytes
   For example, if the multiplex assay panel consists of five analytes (IL-2, IL-3, IL-4, IL-10, and IL-12), the amount of wash buffer needed for each well:
   50 μL – (0.1 μL × 5) = 49.5 μL
4. Gently mix each biotinylated antibody stock vial.
5. Add the required amount of each antibody into the required amount of biotinylated antibody diluent.
   a. Required amount of antibody = 0.1 μL × number of wells
   b. Required volume of biotinylated antibody diluent = result of step 3 × number of wells
   Note: When using a multichannel transfer pipette, 10 additional wells of reagents are required for pipetting loss and 5 wells if a single channel pipette is used.
6. Mix by gentle vortexing.

Protocol 7.6.4: Streptavidin-PE Working Solution Preparation

1. Determine the number of wells needed for the experiment.
2. Calculate the amount of stock streptavidin-PE needed for each well:

The working concentration of streptavidin-PE is 4 μg/mL. The total volume for each well is 50 μL. The stock streptavidin-PE concentration is usually 2 mg/mL, that is, 500× stock. Therefore, for each well, 0.1 μL of streptavidin-PE stock is required.

Note: Concentrations of streptavidin-PE may vary from lot to lot. It will have to be calculated so that the working concentration of streptavidin-PE is 4 μg/mL.

3. Gently mix the streptavidin-PE stock vial.
4. Add the required amount of streptavidin-PE into the required amount of wash buffer.
    a. Required amount of antibody = 0.1 μL × number of wells
    b. Required volume of wash buffer = 50 μL × number of wells

    Note: When using a multichannel transfer pipette, 10 additional wells of reagents are required for pipetting loss and 5 wells if a single channel pipette is used.
5. Mix by vortexing.

Protocol 7.6.5: Performing the Bead Immunoassay

Prepare the 96-well multiscreen microtiter filter plates.

1. Prepare the template. Eight multiplexed standard dilutions for the standard curve. Standard curves and samples should be run in duplicates or triplicates.
2. If the whole plate will not be used, seal the unused well with an aluminum foil lid.

Step-by-Step Protocol

1. Prewet the 96-well filtration plate with wash buffer (100 μL/well).
2. Remove buffer by applying vacuum with the multiscreen vacuum manifold.
3. Add 50 μL of antibody-coupled bead-working suspension to each well.
4. Remove solution by applying vacuum with the multiscreen vacuum manifold. Gently tap the plate bottom onto several layers of paper towels to remove any residual buffer.
5. Add 50 μL of standards or test samples to each well.

    Note: Test sample may have to be diluted with sample diluent. Refer to Note 4 of the Troubleshooting Guide for details.
6. Cover the plate with an aluminum foil lid.
7. Incubate on the shaker for 30 min at room temperature. Protect from light.
8. Remove the aluminum foil lid.
9. Wash the wells three times with 100 μL wash buffer.
10. Add 50 μL of biotinylated antibody working solution to each well.
11. Cover the plate with an aluminum foil lid.
12. Incubate on the shaker for 30 min at room temperature. Protect from light.
13. Remove the aluminum foil lid.
14. Wash the wells three times with 100 μL wash buffer.
15. Add 50 μL of streptavidin-PE working solution to each well.
16. Cover the plate with an aluminum foil lid.

17. Incubate on the shaker for 10 min at room temperature. Protect from light.
18. Remove the aluminum foil lid.
19. Wash the wells three times with 100 μL wash buffer.
20. Add 125 μL of wash buffer to each well to resuspend the beads.
21. Cover the plate with an aluminum foil lid.
22. Place the plate on the microtiter shaker and shake for 30 s at 1100 rpm.
23. Remove the aluminum foil lid.
24. Read samples on a Luminex reader or a flow cytometer.

Troubleshooting Guide

Note 1: Selection of Antibody Pairs

As with other sandwich FIA or ELISA, the key to developing a successful bead-based multiplex assay is a good antibody pair. Therefore, the first step is to obtain optimal antibody pairs by screening different combinations of capture and detection antibodies. We have found that antibody pairs optimized for sandwich ELISA usually work well with bead-based assays. Most bead-based assays can be developed by screening three to five different antibody pairs from vendors that provide antibodies for ELISA. If both capture and detection antibodies are monoclonal, it is important that they recognize two different epitopes of the same analyte, and the two epitopes, if known, should be as far apart as possible to avoid interference. Sometimes, it is easier to have a combination of monoclonal and affinity-purified polyclonal antibodies. Because polyclonal antibodies react against multiple epitopes, they can better serve as capture antibodies in some assays. However, although it may be easier to make assay reagents with monoclonal and polyclonal combinations, the drawbacks of using polyclonal antibodies are the lot-to-lot variation and higher incidences in cross-reactivity, especially in the multiplex assay format.

Note 2: Antibody Specificity

In multiplex immunoassays, specificities of antibodies are crucial; the antibody pair must not cross-react with other analytes and antibodies in the panel. As sandwich FIA relies on two antibodies, one highly specific antibody can sometimes compensate for the other less specific one. It seems that the specificity of the detection antibody is more critical in multiplex immunoassays. One way to test the specificity of the antibody in the multiplex assay is to look at the percentage of signal generated from the assay in the presence of all the other antigens except the one for the antibody pair being tested. For example, in a 3-plex cytokine assay of IL-12, IL-4, and IL-10, cross-reactivity of the IL-12 antibody pair to IL-4 and IL-10 is tested by comparing the following conditions:

1. Full 3-plex assay in the presence of capture and detection antibodies of IL-12, IL-4, and IL-10 and all three recombinant antigens.

2. An assay in the presence of capture and detection antibodies of IL-12, IL-4, and IL-10 and recombinant IL-4 and IL-10 protein standards.

Note 3: Recombinant Proteins

Many protein assays require calibration standard curves to be included in the experiment so that concentrations of analytes in the samples can be calculated according to the assay signal. It is very important to obtain recombinant protein standards that mimic the native proteins in samples such as cell culture supernatant, serum, or plasma. The stability and solubility of the recombinant proteins should also be evaluated. Insolubility of the recombinant protein can have a major impact on the final assay result.

Note 4: Assay Diluents

The assay diluent buffer for generating the standard curve should mimic the conditions in the samples. For example, to assay cell culture supernatant samples, fresh medium with the same supplement serum amount and other factors can be used to dilute the standards. For serum, plasma, and other biological samples such as urine, broncho-alveolar lavage (BAL), or spinal fluid, the diluents for standard and sample have to be optimized for each sample matrix as well as assay panels to minimize factors that interfere with assay signals and dose recoveries, that is, assay accuracies. Similar to other ELISAs, the sample diluent should contain reagents to reduce effects of human anti-animal antibodies such as human anti-mouse antibodies (HAMA) (Klee, 2000; de Jager and Rijkers, 2006) and other factors interfering with analyte measurement (Phillips et al., 2006). The authors also found that it is important to dilute the serum or plasma samples at least 1- to 3- fold to reduce bead aggregations.

The standard diluent should contain the same amount of sample diluent plus protein matrixes to mimic the assay environment to achieve assay accuracy. For cytokine assays, normal-pooled sera or plasmas from several individuals of the same species of the assay samples may be a good diluent to mimic the assay matrixes. However, it may be very difficult to find human "normal" donors with very low (i.e., under assay detection level) endogenous levels of some cytokines in their sera or plasmas. It will be even more difficult if the assays are for determination of other analytes such as hormones or matrix metalloproteinases (MMPs). Therefore, the standard diluents will have to be optimized for different assay panels. Sometimes, commercially available standard and sample diluents for ELISA may be a good choice for the bead-based assays (Pfleger et al., 2008).

Note 5: Secondary Antibody Diluent

The diluent buffer for biotinylated detection antibodies should contain protein-blocking reagents to reduce nonspecific background signal. For example, small amounts (0.5–5%)

of normal mouse serum can serve as a blocking reagent if mouse monoclonal detection antibodies are employed. Concentrations of detection antibodies for the assays should be titrated by testing standard curves and blanks with various dilutions of the detection antibodies. The goal of the titration is to obtain low background signals and optimal signal-to-noise ratios.

Key References

de Jager, W. and Rijkers, G. T. (2006). Solid-phase and bead-based cytokine immunoassay: A comparison. *Methods* 38: 294–303.

Gottschalk, P. G. and Dunn, J. R. (2005). The five parameter logistic: A characterization and comparison with the four-parameter logistic. *Analytical Biochemistry*, 343: 54–65,

Klee, G. G. (2000). Human antimouse antibodies. *Archives of Pathology and Laboratory Medicine*, 124: 921–923.

Pfleger, C. et al. (2008). Effect of serum content and diluent selection on assay sensitivity and signal intensity in multiplex bead-based immunoassays. *Journal of Immunological Methods*, 329: 214–218.

Phillips, D. J. et al. (2006). Interference in microsphere flow cytometric multiplexed immunoassays for human cytokine estimation. *Cytokine*, 36: 180–188.

Skogstrand, K. et al. (2005). Simultaneous measurement of 25 inflammatory markers and neurotrophins in neonatal dried blood spots by immunoassay with xMAP technology. *Clinical Chemistry*, 51: 1854–1866.

Chapter 8

Carbohydrate Arrays

Denong Wang

Contents

Introduction

Carbohydrates are prominently displayed on the surfaces of cells and are present in many secretory proteins in bodily fluids. Expression of complex carbohydrates by human cells is characteristically associated with the stages or steps of embryonic development, cell differentiation, as well as transformation of normal cells to abnormally differentiated tumor or cancer cells.[1–4] Sugar moieties are also abundantly expressed on the outer surfaces of the majority of viral, bacterial, protozoan, and fungal pathogens. Many sugar structures are pathogen specific, which makes them important molecular targets for pathogen recognition, diagnosis of infectious diseases, and vaccine development.[5–10] Exploring the biological information content in diverse sugar chains is one of the current foci of glycomics research and technology development.

In the past few years, a number of experimental approaches to the construction of carbohydrate microarrays have been developed.[11–20] In spite of their technological differences, these carbohydrate microarrays are all solid-phase binding assays for carbohydrates and their interaction with other biological molecules. They share a number of common characteristics and technical advantages. First, they have the capacity to display a large panel of carbohydrates in a limited chip space. Second, each carbohydrate is spotted in an amount that is drastically smaller than that required for a conventional molecular or immunological assay. Thus, the bioarray platform makes effective use of carbohydrate substances. Third, they have high detection sensitivity. The microarray-based assays have higher detection sensitivity than most conventional molecular and immunological assays. This was attributed to the fact that the binding of a molecule in solution phase to an immobilized micro spot of ligand in the solid phase exhibits minimal reduction of the molar concentration of the molecule in solution.[21] Therefore, in a microarray assay, it is much easier for a binding equilibrium to take place, giving high sensitivity.

Carbohydrate microarrays constructed by various methods may differ in their technical features and suitability for a given practical application. For example, the method of nitrocellulose-based immobilization of carbohydrate-containing macromolecules is suitable for the high-throughput construction of carbohydrate antigen microarray.[11,14,18,22,23] This platform of carbohydrate microarrays is readily applicable for the large-scale immunological characterization of carbohydrate antigens and anticarbohydrate antibodies, and interaction of carbohydrates and other receptors. However, the detection specificity of this carbohydrate microarray would be at the level of an antigen, not a glycoepitope, if the native carbohydrate antigens were spotted. This is owing to the fact that many carbohydrate antigens, such as polysaccharides, glycoproteins, and glycolipids, display more than one antigenic determinant, including glyco- or nonglyco- or both epitopes. Examining the finer details of the binding properties would require the use of microarrays of pure saccharide sequences. The mono- and oligosaccharide array-based binding assays can be applied, in combination with saccharide competition assays, to decipher precise saccharide components of a specific antigenic determinant or glycoepitope.[19]

List of Protocols

Protocol Number	Name
8.1	Design and Construction of Sugar Arrays
8.2	Micro Spotting of Carbohydrates onto Bioarray Substrates
8.3	Immunostaining of Carbohydrate Microarrays
8.4	Microarray Scanning and Data Collection
8.5	Microarray Data Processing and Standardization
8.6	Validation and Further Investigation
8.7	Probing Immunogenic Sugar Moieties Using Sugar Arrays

Our laboratory and collaborators have developed two complementary platforms of carbohydrate arrays, that is, carbohydrate antigen arrays.[11,14,18,22,23] and photo-generated glycan arrays.[19,24] The former makes use of a nitrocellulose-based bioarray substrate that supports noncovalent immobilization of carbohydrate-containing macromolecules, including carbohydrate antigens of distinct structural configurations, such as polysaccharides, glycoproteins, and glycolipids; the latter allows covalent coupling of saccharides, including mono-, oligo-, and polysaccharide microarrays, on a photoactive chip substrate.

These technologies take advantage of the existing cDNA microarray system, including spotter and scanner, for efficient production and application of carbohydrate microarrays, as well as bioinformatics tools that facilitate the processing of the large data set produced by each carbohydrate microarray assay. Both approaches are sensitive enough to recognize the profiles of anticarbohydrate antibodies with as little as a few microliters of serum specimen and have the chip capacity to handle the antigenic preparations of most common pathogens (~20,000 micro spots per biochip).

We describe, in the following text, Protocols 8.1–8.7 that support the application of the two platforms of a carbohydrate microarray. We summarize the key steps of carbohydrate microarray applications as (1) design and construction of sugar arrays, (2) micro spotting molecules onto bioarray substrates, (3) immunostaining and

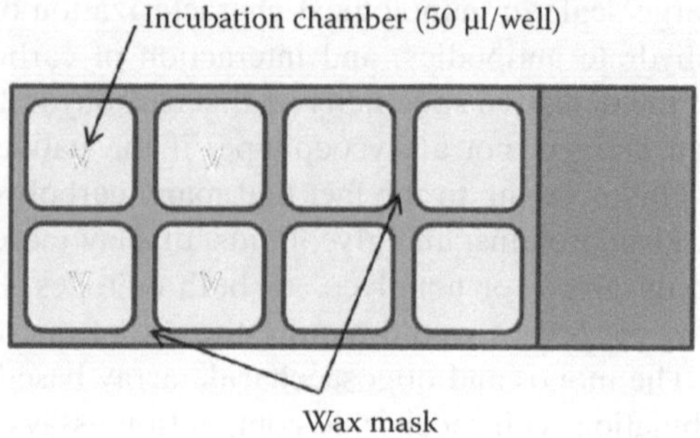

Figure 8.1
An illustration of the eight-chamber subarrays. A microarray slide imprinter (ArrayIt, Sunnyvale, CA) is used to create multichamber subarrays on the surface of a glass slide substrate before or after spotting of carbohydrate microarrays.

scanning of arrays, (4) analysis of microarray data, (5) validation of microarray data using conventional immunological assays, and (6) identification of carbohydrate-based biomarkers using complementary platforms of carbohydrate microarrays.

Protocol 8.1: Design and Construction of Sugar Arrays

One may use the full surface of a microscope slide to construct a "repertory" carbohydrate microarray of ~20,000 spots capacity for biomarker discovery. One may also use a multichamber subarray system to construct customized carbohydrate microarrays for defined purposes. We have been using the latter more often in our laboratory research and clinical applications. An example is illustrated in Figure 8.1. In this case, each glass slide is separated into eight subarrays. The microarray capacity is ~500 micro spots per subarray, with spot sizes of approximately 200 μm and at 300 μm intervals, center to center. A single slide is thus designed to enable eight microarray assays.

Repeats and dilutions: We usually print carbohydrate antigens at the initial concentration of 0.5–1.0 mg/mL. The absolute amount of antigens or antibodies printed on a chip substrate is in the range of 0.5–1.0 ng per micro spot. They are further diluted at 1:3, 1:9, and 1:27 concentrations. A given concentration of each preparation is repeated at least three times to allow statistical analysis of detection of identical preparation at a given antigen concentration.

Antibody isotype standard curves: Antibodies of IgG, IgA, and IgM isotypes of corresponding species are printed at given concentrations to serve as standard curves in microarray format. This design allows quantifying antibody signals that are captured by spotted carbohydrate antigens. In addition, such standard curves are useful for microarray data normalization and cross-chip scaling of microarray detection.

Protocol 8.2: Micro Spotting of Carbohydrates onto Bioarray Substrates

How It Works

Using Cartesian Technologies' (Irvine, CA) PIXSYS 5500C, a high-precision robot designed for cDNA microarrays, carbohydrates of various complexities (see Note 1) are picked up by dipping quill pins into carbohydrate solutions and printed onto bioarray substrates. The complementary AxSys™ software (Cartesian Technologies, Irvine, CA) is used to instruct movement of pins related to the dispense platform and the printing process. For spotting carbohydrate antigen arrays, we use the nitrocellulose-coated glass slides as bioarray substrates. For photo-generated glycan arrays, saccharides are printed onto the phthalimide-amine monolayer (PAM)-coated slides. Carbohydrates are immobilized to nitrocellulose by physical-chemical adsorption involving noncovalent interactions such as H-bonding and ionic and hydrophobic interactions with the substrate. For covalent tethering of polysaccharides, the PAM slide is used. This is accomplished by a free radical mechanism in which a tertiary hydrogen is abstracted from the carbohydrate during free radical formation. The phthalimide group of the monolayer also forms an unstable free radical intermediate under UV irradiation. Carbohydrate radicals and phthalimide radicals that are in close proximity to each other react to form a stable covalent bond. Therefore, UV irradiation of spotted carbohydrate arrays permanently grafts the carbohydrates onto the slide's surface to produce covalently bound carbohydrate microarrays.

Required Materials

Apparatus

Micro spotting: Cartesian Technologies' (Irvine, CA) PIXSYS 5500C
Bioarray substrates:
  FAST slides (Schleicher & Schuell, Keene, NH)
  PAM-coated slides[19,24]
Wax imprinter: Microarray Slide Imprinter (ArrayIt, Sunnyvale, CA)
Array scanning: ScanArray 5000A microarray scanner (PerkinElmer, Torrance, CA)
Multiple-RAY 8 Watt Laboratory lamp, UV lamp, 300 nm excitation (UVP Inc., Upland, CA)
Quartz vials (Corning, Corning, NY)

Softwares

Array design: CloneTracker (Biodiscovery, Marina del Rey, CA)
Array printing: AxSys (Cartesian Technologies, Irvine, CA)
Array scanning and analysis: ScanArray Express (PerkinElmer, Torrance, CA)

Antibodies

Rabbit anti-*Bacillus anthracis* spore polyclonal IgG antibodies (Abcam, Cambridge, U.K.)
Streptavidin-Cy3 and Streptavidin-Cy5 conjugates (Amersham Pharmacia, Piscataway, NJ)

Species-specific anti-immunoglobulin antibodies and their fluorescent conjugates, Cy3, Cy5, or FITC (Sigma, St. Louis, MO; BD-PharMingen, San Diego, CA).

Reagents and Buffers

Dilution buffer: saline (0.9% NaCl)
Rinsing solution: 1× PBS, pH 7.4 w/0.05% (v/v) Tween 20
Blocking solution: 1% (w/v) BSA in PBS w/0.05% (w/v) NaN_3
Argon
Nitrogen

Protocol 8.2.1: Printing of Carbohydrate Arrays onto Nitrocellulose Slides

Step-by-Step Protocol

1. Prepare samples of carbohydrate antigens and antibody standards in 0.9% NaCl at 0.1–0.5 mg/mL, and transfer them in 96-well plates (see Note 2 and Note 3 in the Troubleshooting Guide).
2. Place the 96-well plates containing samples on the Cartesian arrayer robot for printing.
3. Adjust the print program so that carbohydrate antigens and antibodies are printed at spot sizes of ~150 μm and at 375 μm intervals, center to center.
4. Spot each antigen and antibody as triplet replicates in parallel.
5. The printed carbohydrate microarrays are air-dried and stored at room temperature overnight before application (see Note 5).

Protocol 8.2.2: Preparation of Photo-Generated Glycan Arrays on PAM Slides[19,24]

Step-by-Step Protocol

1. Follow Protocol 8.2.1 through step 4 for printing down carbohydrates but using the PAM slides prepared earlier as the substrate.
2. Air-dry the carbohydrates spotted on PAM slides, and then place individual slides in quartz vials.
3. Purge the vial with argon or nitrogen gas, and seal.
4. Place the quartz vial containing the PAM slide under a UV lamp (excitation wavelength, ~300 nm).
5. The printed-array side of the slide should face the lamp.
6. Irradiate each slide for 1 h under the UV lamp in order to photo-cross-link the carbohydrates to the surface. Caution: Avoid eye contact with the lamp during this process. Use protective goggles rated for safe viewing at the 300 nm wavelength.

Protocol 8.3: Immunostaining of Carbohydrate Microarrays

The staining procedure for carbohydrate microarrays is basically identical to the routine procedure for immunohistology. Immunostaining steps of carbohydrate arrays are as follows (see Note 4 and Note 6 in the Troubleshooting Guide):

1. Rinse printed microarray slides with 1× PBS, pH 7.4, with 0.05% Tween 20 for 5 min.
2. Block slides with 1% BSA in PBS containing 0.05% NaN_3 at room temperature for 30 min.
3. Stain each subarray with 50 μL of test sample, which is diluted in 1% BSA PBS containing 0.05% NaN_3 and 0.05% Tween 20.
4. Incubate the slide in a humidified chamber at room temperature for 60 min.
5. Wash slides five times with 1× PBS, pH 7.4, with 0.05% Tween 20.
6. Stain slides with 50 μL of titrated secondary antibodies. Anti-human (or other species) IgG, IgM, or IgA antibodies with distinct fluorescent tags, Cy3, Cy5, or FITC, are mixed and then applied on the chips.
7. Incubate the slide in a humidified chamber with light protection at room temperature for 30 min.
8. Wash slides five times.
9. Place the slide in a 50 mL Falcon™ centrifuge tube, and spin at 1000 rpm for 2 min to remove the washing buffer.
10. Cover the slides in a histology slide box to prevent fluorescent quenching of signal by lights.

Protocol 8.4: Microarray Scanning and Data Collection

Scan the microarray with ScanArray 5000A microarray scanner (PerkinElmer Life Science) following the instructions in the manufacturer's user manual. Fluorescence intensity values for each array spot and its background are calculated using ScanArray Express software. A staining result is considered positive when the mean fluorescent intensity value of micro spot is significantly higher than the mean background of the identically stained microarray with the same fluorescent color (see Note 7 in the Troubleshooting Guide).

Protocol 8.5: Microarray Data Processing and Standardization

A number of advanced software packages are available for microarray data normalization, statistical analysis, and pattern-recognition-based advanced data processing (http://genome-www5.stanford.edu/resources/restech.shtml). We have been using Stanford University's

Significance Analysis of Microarrays (SAM, http://www-stat.stanford.edu/~tibs/SAM/), Prediction analysis for microarrays (PAM http://www-stat.stanford.edu/~tibs/PAM/index.html), and SAS Institute's JMP 6.0—genomics, proteomics, and microarrays software package (http://www.jmp.com/; see Note 8 in the Troubleshooting Guide).

Protocol 8.6: Validation and Further Investigation of Microarray Observations

A microarray finding may require further validation by other experimental approaches. We usually confirm our results by at least one of the alternative immunoassays, such as ELISA, dot blot, Western blot, flow cytometry, and immunohistology. However, the epitopes or antigenic determinants displayed by a carbohydrate antigen in a specific biarray substrate may not be necessarily identical to those that are displayed by other assay systems. Thus, there is a possibility that a "chip hit" is not reproduced by other assays. In such circumstances, one may conduct multiple carbohydrate array assays to confirm the initial microarray observation.

Protocol 8.7: Probing Immunogenic Sugar Moieties Using Sugar Arrays

Different platforms of carbohydrate arrays may be applied complementarily to address a biological question. For example, we have explored the use of two bio-array platforms, carbohydrate antigen arrays, and photo-generated glycan arrays, to probe the potential immunogenic sugar moieties expressed by the spores of *Bacillus anthracis*. Our rationale was that if *B. anthracis* spores expressed antigenic carbohydrate structures, then it would be possible for immunizing animals using the spore preparations to elicit antibodies specific for these structures.[25] This assumption was made on the basis of the fact that the host immune system is able to recognize subtle changes in sugar structures, especially those that are exposed on the surfaces of microbial pathogens that are foreign components of the mammalian hosts.

Using the first technology, we spotted a large panel of carbohydrate antigens in the nitrocellulose-based microarray substrate. Then, we applied these arrays to examine whether rabbit antisera elicited by anthrax spore antigens contain anti-carbohydrate reactivities with certain carbohydrate structures presented by the arrays. We discovered that these rabbit antisera strongly react with a purified polysaccharide of *Streptococcus pneumoniae* type 23 (Pn 23). This carbohydrate antigen displays an L-rhamnose-containing antigenic determinant with strong antigenic cross-reactivities among Pn 23 streptococcal groups B and G.[26] Thus, this microarray analysis provided an important chip hit of a potential immunogenic sugar moiety of *B. anthracis* spores a few years before the structural elucidation of the saccharide moiety displayed by the exosporium BclA glycoprotein.[25]

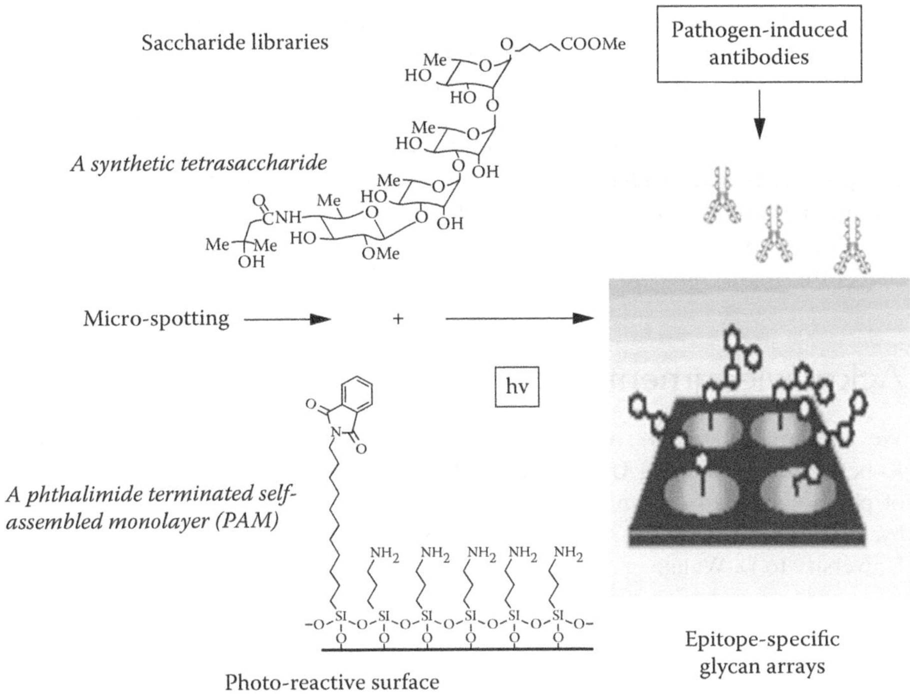

Figure 8.2
Photo-generated glycan arrays for rapid identification of pathogen-specific immunogenic sugar moieties. This diagram shows that a panel of mono-, oligo-, and polysaccharides were spotted on the photoreactive glass slides followed by UV irradiation to induce covalent coupling of the saccharides to the array substrate. To probe potential immunogenic sugar moieties, pathogen-specific antibodies were applied to react with these glycan arrays. A reproducible positive detection in this assay may lead to rapid recognition of pathogen-specific carbohydrate structures of immunological importance. (From Wang, D. et al., *Proteomics* 7: 180–184, 2007. With permission.)

We further characterized the precise carbohydrate structures that are recognized by the rabbit antisera using the photo-generated glycan arrays. This method employs a glass slide coated with a self-assembled mixed monolayer that presents photoactive phthalimide chromophores at the air–monolayer interface. Upon exposure to UV radiation (300 nm), the phthalimide end groups graft the spotted carbohydrates by hydrogen abstraction followed by radical recombination. Thus, a unique technical advantage of this method is the ability to produce epitope-specific glycan arrays using unmodified mono- and oligosaccharides. We applied, therefore, this technology to display a large panel of saccharide structures, including synthetic fragments and derivatives of the anthrose-containing tetrasaccharide side chain of the *B. anthracis* exosporium and a number of control carbohydrate antigens, for an immunological characterization. A schematic overview of this biomarker identification strategy is shown in Figure 8.2.

This glycan array analysis confirmed that a tetrasaccharide of BclA glycoprotein bares a dominant antigenic determinant, which is composed of a terminal anthrose residue and three adjacent L-rhamnoses. The terminal trisaccharide unit is essential for the constitution of a highly specific antigenic determinant. Given the fact that this carbohydrate moiety is displayed on the outermost surfaces of *B. anthracis* spores and its expression is highly specific for the spore of *B. anthracis*, the anthrose-containing tetrasaccharide can be considered an important immunological target. Its applications may include identification of the presence of *B. anthracis* spores, surveillance and diagnosis of anthrax infection, and development of novel vaccines targeting the *B. anthracis* spore. Efforts must also be made to explore the biological role of this highly specific carbohydrate moiety of *B. anthracis*.

Acknowledgment

We are grateful to Drs. Gregory T. Carroll, Nicholas J. Turro, and Jeffrey T. Koberstein of Columbia University for their contribution to the joint development of photo-generated glycan arrays. This work is currently supported by, or in part by, NIH grants U01CA128416, HL084318-01A1, and AI064104 from Stanford University to D. Wang.

Key References

Carroll, G. T. et al. (2006). Photochemical micropatterning of carbohydrates on a surface. *Langmuir*, 22: 2899–2905.

Wang, D. et al. (2002). Carbohydrate microarrays for the recognition of cross-reactive molecular markers of microbes and host cells, *Nature Biotechnology*, 20: 275–281.

Wang, D. et al. (2007). Photogenerated glycan arrays identify immunogenic sugar moieties of *Bacillus anthracis* exosporium. *Proteomics*, 7: 180–184.

Wang, D. and Lu, J. (2004). Glycan arrays lead to the discovery of autoimmunogenic activity of SARS-CoV. *Physiological Genomics*, 18: 245–248.

Troubleshooting Guide

Note 1: Antigen Preparations Suitable for the Nitrocellulose Bioarray Substrate

Carbohydrate antigen of multiple structural configurations, including polysaccharides, natural glycoconjugates, oligosaccharide–protein and oligosaccharide–lipid conjugates, are applicable for platform of carbohydrate antigen microarrays.[11,14,18] In addition to printing carbohydrate microarrays, this platform is also applicable for producing protein microarrays.[22] As with carbohydrate microarrays, there is no need to chemically conjugate a protein for its surface immobilization. However, it is recommended that each antigen preparation be tested on a chip substrate for the efficacy of immobilization and expression of antigenic determinants.

Note 2: Preservation of Polysaccharides

Purified polysaccharides are generally stored as dried powder at room temperature. They can also be preserved in saline solutions (0.9% NaCl) containing a droplet of chloroform and stored at 4°C for a long period of time.

Note 3: Printing of Samples

Before loading sample solutions onto the arrayer, it is important to spin the solution in an Eppendorf centrifuge at maximum speed (at least 15,000× g) for at least 15 min. Before and after each arraying experiment, it is recommended to examine and clean the printing pins.

A test slide is usually used to optimize the quality of printing. The water supply of the Cartesian Arrayer should constantly be checked during the arraying experiment to ensure adequate flow to the wash chamber.

Note 4: Examination of Presence of Samples on Array and Their Antigenic Structures

It is essential to examine whether proteins, synthetic peptides, and carbohydrates are successfully printed and whether desired epitopes or antigenic determinants are preserved on the chips. The printed microarrays can be incubated with antibodies, receptors, or lectins known to react with the printed substance. The reaction is detected either by conjugating a fluorochrome to the "detector" directly, or by a second-step staining procedure.

Note 5: Storage of Printed Carbohydrate and Protein Microarrays

The arrays are usually air-dried and stored at room temperature. For long-term preservation, the chips can be sealed in a plastic bag with desiccant and stored at –20°C.

Note 6: Staining Considerations

After the last wash between each staining step, it is important to completely withdraw the wash buffer inside reaction chambers. Otherwise, the remaining buffer may lower the antibodies concentration to be analyzed.

Note 7: Scanning and Data Collection

Training with the technical experts of PerkinElmer is necessary before performing microarray scanning and data collection using the ScanArray Express software package.

Note 8: Microarray Data Analysis and Standardization

It is important to conceptually understand the functions of the bioinformatics tools in use in order to correctly interpret the results. Our current general approach to the standardization of carbohydrate microarray data sets before clinical statistical analysis is summarized as follows. Carbohydrate array data sets are preprocessed and statistically analyzed using JMP Microarray software from SAS Institute. Antigen-specific antibody reactivity is shown as microarray scores, which are the log 2-transformed microarray values mean and mean background. Then, we utilized an antigen-by-antigen ANOVA model to obtain statistically significant differences. Data from triplicate spots for each antigen were included in the ANOVA model for that antigen. A cutoff to detect significant differences is determined by applying a multiple testing correction to statistical results from the ANOVA model.

Note 9: Biosafety Procedures

When working with chemicals, it is advisable to wear suitable protective clothes, such as laboratory coat and disposable gloves. When human serum specimens are involved, experiments must be conducted in accordance with the standard biosafety procedures instituted by the CDC and WHO.

References

1. Feizi, T. (1982). The antigens Ii, SSEA-1 and ABH are in interrelated system of carbohydrate differentiation antigens expressed on glycosphingolipids and glycoproteins. *Advances in Experimental Medicine and Biology*, 152: 167–177.
2. Hakomori, S. (1985). Aberrant glycosylation in cancer cell membranes as focused on glycolipids: overview and perspectives. *Cancer Research*, 45: 2405–2414.
3. Focarelli, R. et al. (2001). Carbohydrate-mediated sperm-egg interaction and species specificity: a clue from the *Unio elongatulus* model. *Cells Tissues Organs*, 168: 76–81.
4. Crocker, P. R. and Feizi, T. (1996). Carbohydrate recognition systems: functional triads in cell-cell interactions. *Current Opinion in Structural Biology*, 6: 679–691.
5. Heidelberger, M. and Avery, O. T. (1923). The soluble specific substance of pneumococcus. *Journal of Experimental Medicine*, 38: 73–80.
6. Dochez, A. R. and Avery, O. T. (1917). The elaboration of specific soluble substance by pneumococcus during growth. *Journal of Experimental Medicine*, 26: 477–493.
7. Ezzell, J. W., Jr. et al. (1990). Identification of *Bacillus anthracis* by using monoclonal antibody to cell wall galactose-N-acetylglucosamine polysaccharide. *Journal of Clinical Microbiology*, 28: 223–231.
8. Robbins, J. B. and Schneerson, R. (1990). Polysaccharide-protein conjugates: a new generation of vaccines. *Journal of Infectious Diseases*, 161: 821–832.
9. Mond, J. J. et al. (1995). T cell-independent antigens type 2. *Annual Review of Immunology*, 655–692.
10. Wang, D. and Kabat, E. A. (1996). Carbohydrate antigens (polysaccharides). In *Structure of Antigens* (Ed. MH. V. V. Regenmortal). CRC Press, Boca Raton, FL, 247–276.

11. Wang, D. et al. (2002). Carbohydrate microarrays for the recognition of cross-reactive molecular markers of microbes and host cells. *Nature Biotechnology*, 20: 275–281.
12. Willats, W. G. T. et al. (2002). Sugar-coated microarrays: A novel slide surface for the high-throughput analysis of glycans. *Proteomics*, 2: 1666–1671.
13. Fazio, F. et al. (2002). Synthesis of sugar arrays in microtiter plate. *Journal of the American Chemical Society,* 124: 14397–14402.
14. Fukui, S. et al. (2002). Oligosaccharide microarrays for high-throughput detection and specificity assignments of carbohydrate-protein interactions. *Nature Biotechnology*, 20: 1011–1017.
15. Houseman, B. T. and Mrksich, M. (2002). Carbohydrate arrays for the evaluation of protein binding and enzymatic modification. *Chemistry and Biology*, 9: 443–454.
16. Park, S. and Shin, I. (2002). Fabrication of carbohydrate chips for studying protein-carbohydrate interactions. *Angewandte Chemie International Edition in English*, 41: 3180–3182.
17. Adams, E. W. et al. (2004). Oligosaccharide and glycoprotein microarrays as tools in HIV glycobiology; glycan-dependent gp120/protein interactions. *Chemistry and Biology*, 11: 875–881.
18. Wang, D. and Lu, J. (2004). Glycan arrays lead to the discovery of autoimmunogenic activity of SARS-CoV. *Physiological Genomics*, 18: 245–248.
19. Wang, D. et al. (2007). Photogenerated glycan arrays identify immunogenic sugar moieties of *Bacillus anthracis* exosporium. *Proteomics*, 7: 180–184.
20. Zhou, X. and Zhou, J. (2006). Oligosaccharide microarrays fabricated on aminooxyacetyl functionalized glass surface for characterization of carbohydrate-protein interaction. *Biosensors and Bioelectronics*, 21: 1451–1458.
21. Ekins, R. et al. (1990). Multispot, multianalyte, immunoassay. *Annales de Biologie Cliniques*, 48: 655–666.
22. Wang, D. (2003). Carbohydrate microarrays. *Proteomics*, 3: 2167–2175.
23. Wang, D. (2004). Carbohydrate antigens. In *Encyclopedia of Molecular Cell Biology and Molecular Medicine II* (Ed. Robert A. Meyers). Wiley-VCH, Weinheim, 277–301.
24. Carroll, G. W. et al. (2006). Photochemical micropatterning of carbohydrates on a surface. *Langmuir*, 22: 2899–2905.
25. Daubenspeck, J. M. et al. (2004). Novel oligosaccharide side chains of the collagen-like region of BclA, the major glycoprotein of the *Bacillus anthracis* exosporium. *Journal of Biological Chemistry*, 279: 30945–30953.
26. Heidelberger, M. et al. (1967). Cross-reactions of the group-specific polysaccharides of streptococcal groups B and G in anti-pneumococcal sera with especial reference to type 23 and its determinants. *Journal of Immunology,* 99: 794–796.

Chapter 9

Lectin Microarrays

Masao Yamada

Contents

Introduction

The use of lectins as capture reagents for the assessment of lectin–glycan interactions is widely recognized as an important frontier in the understanding of various aspects of biological phenomena. This chapter provides an introduction to a lectin microarray system jointly developed by Moritex Corporation and AIST (National Institute of Advanced Industrial Science and Technology), both from Japan.

It is well known that (1) glycans stabilize and modify the biological functions of proteins through a process known as *glycosylation*, and over 50% of human proteins are glycosylated; (2) glycan structures change depending on cell differentiation, disease, and overall health condition; (3) glycan structures differ in cell types and organs; (4) a number of immunological activities are initiated through recognition of cell-surface glycan structures expressed on lymphocytes by glycan-binding proteins (i.e., lectins, etc.); and (5) influenza virus infection is initiated through recognition of a specific glycan structure expressed on tracheal cell surfaces by influenza hemagglutinins. These are only a few examples.

It is not an exaggeration to say that without the glycosylation process, life-forms could not exist. For this reason, it is also often said that glycans are the third tier of bioinformative macromolecules after nucleic acids and proteins. Despite this apparent importance of glycans, it is a fact that there are still many undiscovered and little-understood lectins within organisms. To this day, a wide variety of unknown glycans mediate biological activities through interaction with unknown lectins.

There are basically three major branches in glycomics research: (1) glycan synthesis, (2) glycan structural analysis, and (3) glycan functional analysis. Glycan structural analysis is considered the first priority as it elucidates the biological functions of the glycans. This fact is easier to understand if you remember that the development of DNA sequencers greatly accelerated the study of genomics. However, the word *sequencer* as a structural analyzer of glycans may not be fully appropriate, because glycans are not linear-chain molecules like nucleic acids and proteins, but have a variety of branching molecular structures.

Our strategy in glycomics takes a different approach from that used in genomics. We have adopted the use of lectins as so-called "biological decipherers" of glycan structures, capitalizing on their binding specificity to glycan structures. A microarray strategy allows us to utilize a number of lectins with different specificity, thereby enabling analyses of small sample amounts with high sensitivity and high throughput.[1]

Mass spectroscopy (MS) is one possible alternative approach used in glycomic structural analysis. It potentially is a very powerful tool; however, it is not versatile enough to be used in structural analysis. For example, MS is not suitable for differentiation of isomers; analysis of O-glycans and analysis of crude samples cannot be performed with MS. It typically requires fairly large sample sizes (>100 μg of proteins) due to a sensitivity issue. Additionally, using MS for glycan structural analysis is not convenient given that pretreatment of the samples is time consuming and bothersome. The required pretreatments include the isolation of glycoproteins and cleaving of the glycans from the core proteins. Note that it is not our intention to say that the lectin microarray methodology is far more powerful than MS; however, we believe that the lectin microarray methodology is potentially very powerful from the following viewpoints: (1) it is quick and easy to use (i.e., structural profiling can be performed without cleavage of glycans from proteins, in other words, directly from fluorescence-labeled glycoproteins), (2) it is highly sensitive (i.e., 100 pg–100 ng order of glycoproteins), and (3) it is high throughput. One thing to note is that the inference performance of the lectin microarray approach may be weaker than that of MS.

One of the most noteworthy features of basic glycomics research is the weak interaction of glycans and lectins (by more than two orders of magnitude) compared with that of antibodies and antigens as well as that of DNA molecules. This fact inevitably created the need for a new technology that is capable of detecting very weak glycan–lectin interactions directly from a liquid phase without any washing process. You may already know that washing is an indispensable process in DNA microarrays and ELISA assays, to remove nonspecific bindings and extra amounts of labeled molecules floating in the liquid phase. Therefore, you are probably wondering why it is preferable to avoid washing in glycomics. The ability to detect these weak glycan–lectin interactions without washing is important because washing can easily break the already weak glycan–lectin binding. The Moritex GlycoStation™ (Moritex Corporation, Yokohama, Japan) is, as far as we know, the only commercially available system that can successfully detect very weak glycan–lectin interactions from a liquid phase without washing.

The Moritex GlycoStation and LecChip™ (lectin microarray) technologies were thus developed according to the previously mentioned basic concept. The GlycoStation shows its true worth in differential profiling, in which a number of glycoforms are compared with one another in a high-throughput fashion with high sensitivity. For instance, GlycoStation can be used for screening and understanding how partial glycan structures on cell surfaces change depending on the condition of the cells (i.e., cancer, disease, differentiation stage, and so on), and to elucidate the roles glycans play and to understand the correlation between glycoforms and cell conditions. In this manner, GlycoStation will contribute to the discovery of new scientific knowledge in the field of glycobiology and glycomics.

Protocol 9.0: Glycan Profiling Using Lectin Arrays

How It Works

The LecChip Lectin Array

The LecChip Ver.1.0 produced by Moritex Corporation of Japan contains 45 lectins (Figure 9.1a). The 45 lectins (see Note 1) were selected from a number of candidates, taking into consideration each lectin's specificity to glycan structures and stability (Table 9.1). The characteristics of lectins as a recognizer of glycans are classified into several categories, for instance, fucose recognizer (PSA, etc.), sialic acid (MAL, etc.), lactose (ECA, etc.), asialo (PHA-E, etc.), poly-lactosamin (LEL, etc.), mannose (ConA, etc.), O-glycan

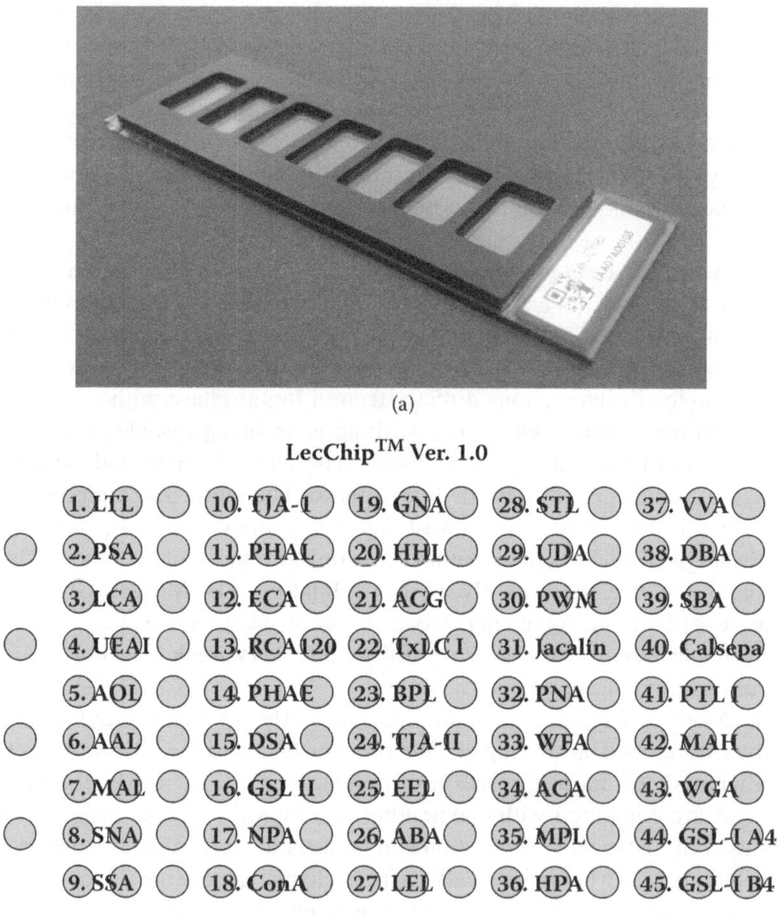

Figure 9.1
(a) The LecChip^{TM} Ver.1.0. (b) Lectin map showing location of immobilized lectins.

TABLE 9.1
Lectin Specificity to Glycan Structures and Stability

Lectin No.	Lectin	Reported Specificity
1	LTL	Fucα1-3 (Galβ1-4)GlcNAc, Fucα1-2Galβ1-4GlcNAc
2	PSA	Fucα1-6GlcNAc, α-D-Glc, α-D-Man
3	LCA	Fucα1-6GlcNAc, α-D-Glc, α-D-Man
4	UEA-I	Fucα1-2Galβ1-4GlcNAc
5	AOL	Fucα1-6GlcNAc (core fucose)
6	AAL	Fucα1-6GlcNAc, Fucα1-3(Galβ1-4)GlcNAc
7	MAL	Siaα2-3Galβ1-4GlcNAc
8	SNA	Siaα2-6Gal/GalNAc
9	SSA	Siaα2-6Gal/GalNAc
10	TJA-I	Siaα2-6Gal/GalNAc
11	PHAL	Tri/tetra-antennary complex-type N-glycan
12	ECA	Galβ1-4GlcNAc
13	RCA120	Galβ1-4GlcNAc
14	PHAE	Bi-antennary complex-type N-glycan with outer Gal and bisecting GlcNAc
15	DSA	(GlcNAcβ1-4)n, Galβ1-4GlcNAc
16	GSL-II	Agalactosylated tri/tetra antennary glycans, GlcNAc
17	NPA	High-Mannose, Manα1-6Man
18	ConA	High-Mannose, Manα1-6(Manα1-3)Man
19	GNA	High-Mannose, Manα1-3Man
20	HHL	High-Mannose, Manα1-3Man, Manα1-6Man
21	ACG	Siaα2-3Galβ1-4GlcNAc
22	TxLC-I	Manα1-3(Manα1-6)Man, bi- and tri-antennary complex-type *N*-glycan, GalNAc
23	BPL	Galβ1-3GalNAc, GalNAc
24	TJA-II	Fucα1-2Galβ1-> or GalNAcβ1-> groups at their nonreducing terminals
25	EEL	Blood group B antigen, Galα1-3Gal
26	ABA	Galβ1-3GalNAc
27	LEL	GlcNAc trimers/tetramers
28	STL	GlcNAc oligomers, oligosaccharide containing GlcNAc and MurNAc
29	UDA	GlcNAcβ1-4GlcNAc, Mixture of Man5 to Man9
30	PWM	(GlcNAcβ1-4)n
31	Jacalin	Galβ1-3GalNAc, GalNAc
32	PNA	Galβ1-3GalNAc
33	WFA	GalNAcβ1-4GlcNAc, Galβ1-3(-6)GalNAc
34	ACA	Galβ1-3GalNAc
35	MPA	Galβ1-3GalNAc, GalNAc
36	HPA	α-Linked terminal GalNAc
37	VVA	α-Linked terminal GalNAc, GalNAcα1-3Gal
38	DBA	blood group A antigen, GalNAcα1-3GalNAc
39	SBA	α- or β-Linked terminal GalNAc, GalNAcα1-3Gal
40	Calsepa	Mannose, Maltose
41	PTL-I	α-Linked terminal GalNAc
42	MAH	Siaα2-3Galβ1-3(Sia α2-6)GalNAc
43	WGA	Chitin oligomers, Sia
44	GSL-I A4	α-Linked GalNAc
45	GSL-I B4	α-Linked Gal

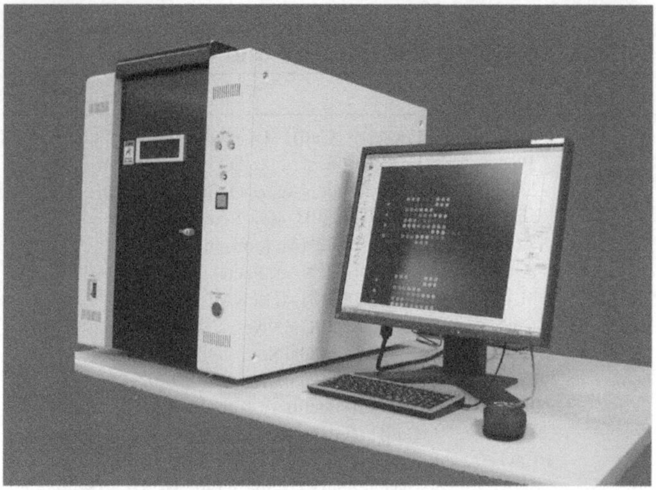

Figure 9.2
The GlycoStation™ Reader 1200.

(Jacalin, etc.), and others. LecChip Ver.1.0 has 7 wells (100 μL well volume) on a glass slide with 45 lectins, each printed in triplicate in each well (Figure 9.1b).

The GlycoStation Reader

The GlycoStation Reader 1200 (see Figure 9.2), a product of Moritex Corporation of Japan, is an ultra-high-speed and highly sensitive scanner based on the principle of evanescent-field fluorescence excitation. The evanescent-field fluorescent excitation method adopted in this reader is essential to detect very weak molecular interactions. If a washing process is applied to the lectin microarray in order to remove redundant proteins that have no binding to lectins, a great deal of affinity information is lost. With a total internal reflection mode, an evanescent field is formed on the surface of the lower-refractive-index side. The penetration depth of the evanescent field is within wavelength distance from the surface, and the field strength decreases exponentially as the distance away from the surface increases. Because of this nature, floating glycans in a liquid phase will show low excitation by the evanescent field, but interacting glycans with lectin probes can be excited effectively. For this reason, one can monitor very weak molecular interactions directly from a liquid phase without any washing (see Note 2).

List of Protocols

The following are the recommended protocols for glycan profiling experiments using the GlycoStation system. Depending on the experiment, there may be a variety of different sample forms, for example, cultivated cells, serum, urine, and other forms of biological fluids. Pretreatment procedures can vary from case to case; however, the core experimental protocol is common to all sample forms. The common denominator is the following process: protein quantification, Cy-3 labeling of glycoproteins, removal of excess free-Cy3 using gel filtration, and ending with the application of Cy-3-labeled glycoproteins onto a LecChip.

In this document, a recommended pretreatment procedure for cultivated cells is described. This is in addition to the preceding standard experimental protocol common to all sample forms.

Protocol Number	Name
9.1	Extraction of Glycoproteins from Cultivated Cells
9.2	Quantification of Proteins
9.3	Cy3 Labeling
9.4	Gel Filtration to Remove Excess Free-Cy3
9.5	Applying Samples to a LecChip
9.6	Scanning the LecChip with the GlycoStation Reader
9.7	Examples—Differential Profiling of Lec1 Mutants

Protocol 9.1: Extraction of Glycoproteins from Cultivated Cells

How it Works

Contained herein is a detailed procedure for extracting glycoprotein samples from cultivated cells and/or from the culture supernatant. With regard to the extraction of glycoproteins, two cases are described. One concerns whole-cell lysates, and the other is a case of fractionating cell cytoplasm, cell membrane proteins, etc.

Recommended cell quantity is in the range of 10^5 to 10^6. Basically speaking, if you have the cell quantity of 1×10^6, the following protocol works for any cell types. However, depending on sample conditions, the order of 10^4 cells will work.

Required Materials

Apparatus

1. Centrifuge (Beckman Coulter, Fullerton, CA, Allegra® X-15R)
2. Centrifuge (TOMY Corp., Tokyo, Japan, MX-301)
3. Sonicator (Bransonic, Danbury, CT, Tabletop Ultrasonic Cleaner, Model 1510)
4. Spin filters (5 kDa MWCO Millipore, Billerica, MA, Cat. No. UFC900596 24pk)
5. Rotator (Nissin Corp., Tokyo, Japan, Rotary Mixer Cat. No. NRC20D, tapping function Cat. No. A-1.5)
6. 96-Well microplate reader (MDS Technologies, SpectraMax M5)
7. LecChips™ (Moritex Corp., Yokohama, Japan)
8. GlycoStation™ Reader (Moritex Corp., Yokohama, Japan)
9. Array-Pro Analyzer Ver. 4.5 (Media Cybernetics, Inc.)

10. GlycoStation™ Tools (Moritex Corp., Yokohama, Japan)
11. Centrifuge desiccator and vacuum unit (EYELA: Tokyo Rikakikai Co., Ltd., Tokyo, Japan, Cat. No. CVE-200D, UT-1000)

Reagents and Buffers

1. PBS: 0.01 M phosphate-buffered saline, pH 7.2–7.4 (Wako Pure Chemical Industries, Ltd., PBS(–)Powder (0.01 mol/L, pH 7.2–7.4), Cat. No. 162-19321)
2. PBS-T: PBS containing 1% Triton X-100 (v/v)
3. TBS: 25 mM Tris, 140 mM NaCl, 3 mM KCl, pH 7.5
4. ProteoExtract Subcellular Proteome Extraction Kit (S-PEK) (Merck, Whitehouse Station, NJ, Cat. No. 539790, contains extraction buffers #1 and #2; protease inhibitor cocktail)
5. Micro BCA™ Protein Assay Reagent Kit (Pierce, Rockford, IL, Cat. No. 23235)
6. Cy3 Mono-Reactive dye pack 1 mg × 5 (GE, Cat. No. PA23001) (see Note 3)
7. DMF: anhydration
8. Zeba™ Desalt Spin Columns, 0.5 ml (25 columns) (Pierce, Rockford, IL, Cat. No. 89882)
9. Probing solution (Moritex Corp., Yokohama, Japan)
10. Kimtex Quarterfold (Kimberly Clark, Dallas, TX, Cat. No. 33560)

Protocol 9.1.1: Cell Pellets

Step-by-Step Protocol

Cells are washed with PBS several times to remove serum-derived glycoproteins contained in cell culture media. The cell pellet is frozen at −80°C.

1. Collect cells from culture by centrifugation at low speed (~200–500× g, 5–10 min) into a cell pellet.
2. Save the cell-free supernatant for Protocol 9.1.4.
3. Resuspend the cell pellet in PBS by gentle vortexing action or by pipetting up and down several times.
4. Centrifuge the cell suspension to reform the pellet.
5. Discard the supernatant.
6. Resuspend the pellet in fresh PBS.
7. Repeat steps (2–4) at least an additional two times.
8. Freeze the washed pellet on dry ice.
9. Store at −80°C prior to use in Protocols 9.1.2 and 9.1.3.

Protocol 9.1.2: Extraction from Whole-Cell Lysate

Step-by-Step Protocol

1. Melt the cell pellet on ice, assuming that a cell pellet is frozen at −80°C.
2. Add 1 mL PBS-T to the cell pellet, and resuspend with a pipette.

3. Sonicate for 1 to 2 min for cell breakage.
4. Centrifuge at 14000× g for 5 min.
5. Recover supernatant with a pipette.
6. Use the recovered protein supernatant as soon as possible to avoid any degradation. However, if you are going to use it afterward, store it at −20°C until needed or, for longer-term storage, freeze at −80°C.

Protocol 9.1.3: Fractionation of Cell Cytosolic, Membrane/Organelle, Nucleic, and Cytoskeletal Proteins

Step-by-Step Protocol

1. Melt a cell pellet contained in a 15 mL tube on ice.
2. Mix 1 mL extraction buffer #1 and 5 μL protease inhibitor cocktail.
3. Add the mixture into the previously mentioned cell pellet tube.
4. Incubate them for 10 min at 4°C with gentle agitation on a rotator.
5. Centrifuge at 1000 × g for 10 min.
6. Transfer the supernatant to a clean tube. This is a *cytosolic* fraction.
7. Mix 1 mL extraction buffer #2 and 5 μL protease inhibitor cocktail.
8. Add the mixture to the 15 mL tube containing the cytosolic fraction.
9. Incubate for 30 min at 4°C with gentle agitation on a rotator.
10. Centrifuge at 6000× g for 10 min.
11. Transfer the supernatant to a clean tube. This is the *membrane/organelle* protein fraction.
12. Use the recovered protein supernatant as soon as possible to avoid any degradation. However, if you are going to use it afterward, store it at −20°C until needed or, for longer-term storage, freeze at −80°C.

Protocol 9.1.4: Extraction of Glycoproteins from Culture Supernatant

Step-by-Step Protocol

In order to analyze glycoproteins contained in culture supernatant, it is necessary to exchange the supernatant with an amino acid-free buffer. This is because the Cy3-labeling reagent reacts with the amino group in the medium, resulting in low efficiency of glycoprotein labeling. Additionally, it is better to concentrate proteins in the buffer in advance because the protein concentration in culture supernatant is usually not so high. Note that this application is limited in a case of serum-free culture medium.

1. Apply 15 mL cell culture supernatant to a 5 kDa cutoff spin filter in a 50 mL tube.
2. Centrifuge (4000× g; 15–45 min) at 4°C in order to concentrate the sample down to about 500 μL volume.
3. Discard the filtrate.

4. Add 15 mL PBS to the concentrated supernatant, and recentrifuge to 500 μL.
5. Discard the filtrate.
6. Add 15 mL PBS to the concentrated supernatant, and recentrifuge to 250 μL.
7. Discard the filtrate.
8. Add PBS to the supernatant up to 500 μL.
9. Recover the supernatant from the filter.
10. Use the recovered protein supernatant as soon as possible to avoid any degradation. However, if you are going to use it afterward, store it at −20°C until needed or, for longer-term storage, freeze at −80°C.

Protocols Common to All Samples

Here is a description for protocols common to all sample forms. The required protein volume is 500 μL with a concentration of 50 μg/mL (25 μg glycoprotein). This value is determined by the quantification process of proteins rather than being a sensitivity issue of the LecChip. As described, the LecChip Ver. 1.0 has 7 wells, each with a volume of 100 μL. Assuming a dilution series starting from a concentration of 2 μg/mL and ending with a concentration of 31.25 ng/mL, the total amount of glycoprotein used is only about 400 ng.

See Protocol 9.1 for required materials for use with these protocols.

Protocol 9.2: Quantification of Proteins

Step-by-Step Protocol

1. Prepare 1.5 mL microcentrifuge tubes labeled as BSA in a tube rack, and add 240 μL PBS to one of the BSA tubes.
2. Add 10 μL of 2 μg/mL BSA solution to the BSA tube for preparing 80 μg/mL BSA solution, and mix it.
3. Place the tube of BSA 80 μg/mL and a tube of the glycoprotein sample prepared from Protocols 9.1.2 to 9.1.4 (earlier).
4. Prepare a serial dilution series of BSA solution starting from a concentration of 80 μg/mL (well volume, 100 μL) in a 96-well microplate, by diluting it by a ratio of 1/2, down to 1/128 of the initial concentration.
5. Prepare the same dilution series in the microplate for the glycoprotein sample.
6. Mix Micro BCA Reagent A (MA) 2.5 mL, Micro BCA Reagent B (MB) 2.4 mL, and Micro BCA Reagent C (MC) 100 μL in a 15 mL tube (hereinafter WR means a working reagent prepared in this way).
7. Add 100 μL WR to all the wells of the microplate.
8. Seal all wells with a plate seal.
9. Incubate at 37°C for 2 h.
10. Remove the plate seal.
11. Measure the absorbance at 562 nm in a microplate reader.

Protocol 9.3: Cy3 Labeling

Step-by-Step Protocol

1. Dilute samples to 50 µg/mL by adding PBS based on the result of the Micro BCA protein assay. In the case of whole-cell lysates, use PBS-T in place of PBS.
2. Add 20 µL of each diluted sample into a tube containing Cy3 amount, which is used for labeling of 100 µg proteins (see Note 3).
3. Thoroughly mix each sample using a pipette.
4. Centrifuge to spin down any fluids on the tube sidewall.
5. Place the tubes in a container shielded from the light, and incubate for 1 h at 25°C.

Protocol 9.4: Gel Filtration to Remove Excess Free-Cy3

Step-by-Step Protocol

1. Prepare two Zeba spin columns.
2. Remove the column's bottom plugs, and loosen caps (do not remove the caps).
3. Place the column in a 2.0 mL microcentrifuge tube.
4. Centrifuge at 1500× g for 1 min to remove storage solution.
5. Mark on the side of the column where the compacted resin is slanted upward. Place the column in the microcentrifuge with the mark facing outward in all the subsequent centrifugation steps.
6. Add 300 µL of TBS on top of the resin bed. Centrifuge at 1500× g for 1 min to remove buffer.
7. Repeat steps 4–6 twice (three times total), discarding buffer from the collection tube.
8. Place the column in a new collection tube, remove the cap, and apply all of the sample (20 µL) to the top of the compact resin bed.
9. Apply 25 µL of TBS to the top of the gel bed after the sample has fully absorbed.
10. Centrifuge at 1500× g for 2 min to collect the sample.
11. Discard desalting columns after use.

Protocol 9.5: Applying Samples to a LecChip™

Step-by-Step Protocol

1. Measure each volume of the Cy3-labeled samples with a micropipette.
2. Prepare a total volume of 1 mL by adding Moritex's probing solution. The protein concentration becomes 1 µg/mL as 1 µg protein is eluted.
3. Serially dilute each sample eluate to 500, 250, 125, 62.5, 31.25, and 15.625 ng/mL using probing solution.

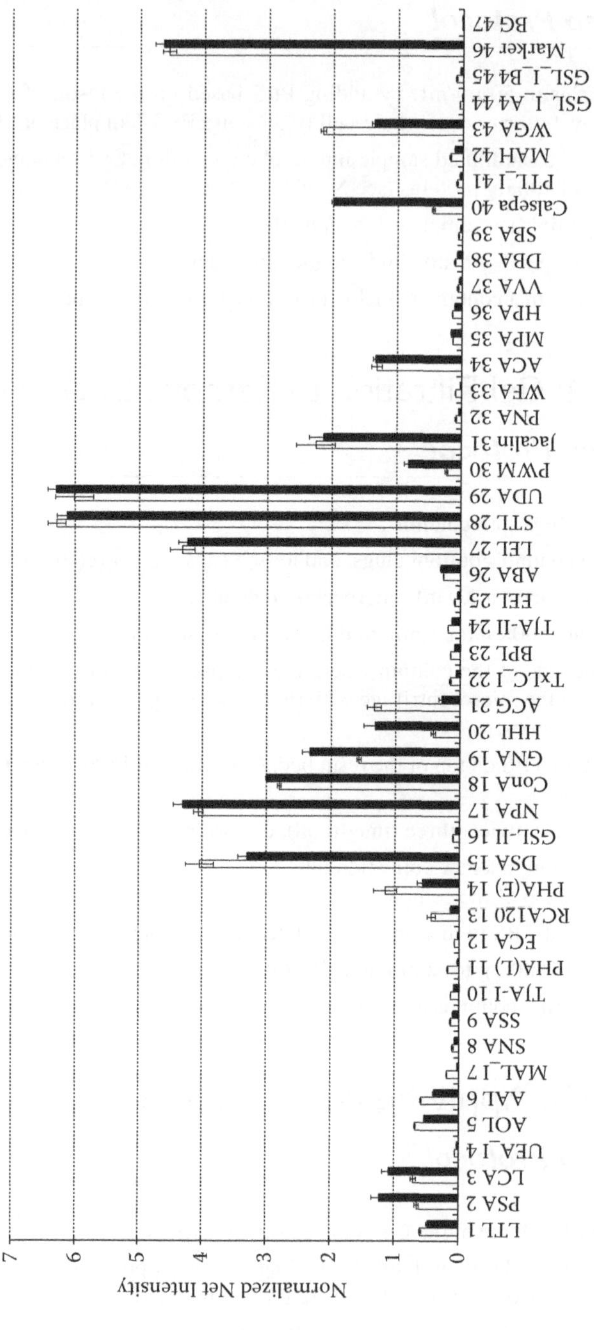
Normalized Net Intensity
0
1
2
3
4
5
6
7
LTL 1
PSA 2
LCA 3
UEA_I 4
AOL 5
AAL 6
MAL_I 7
SNA 8
SSA 9
TJA-I 10
PHA(L) 11
ECA 12
RCA120 13
PHA(E) 14
DSA 15
GSL-II 16
NPA 17
ConA 18
GNA 19
HHL 20
ACG 21
TxLC_I 22
BPL 23
TJA-II 24
EEL 25
ABA 26
LEL 27
STL 28
UDA 29
PWM 30
Jacalin 31
PNA 32
WFA 33
ACA 34
MPA 35
HPA 36
VVA 37
DBA 38
SBA 39
Calsepa 40
PTL_I 41
MAH 42
WGA 43
GSL_I_A4 44
GSL_I_B4 45
Marker 46
BG 47
(a)

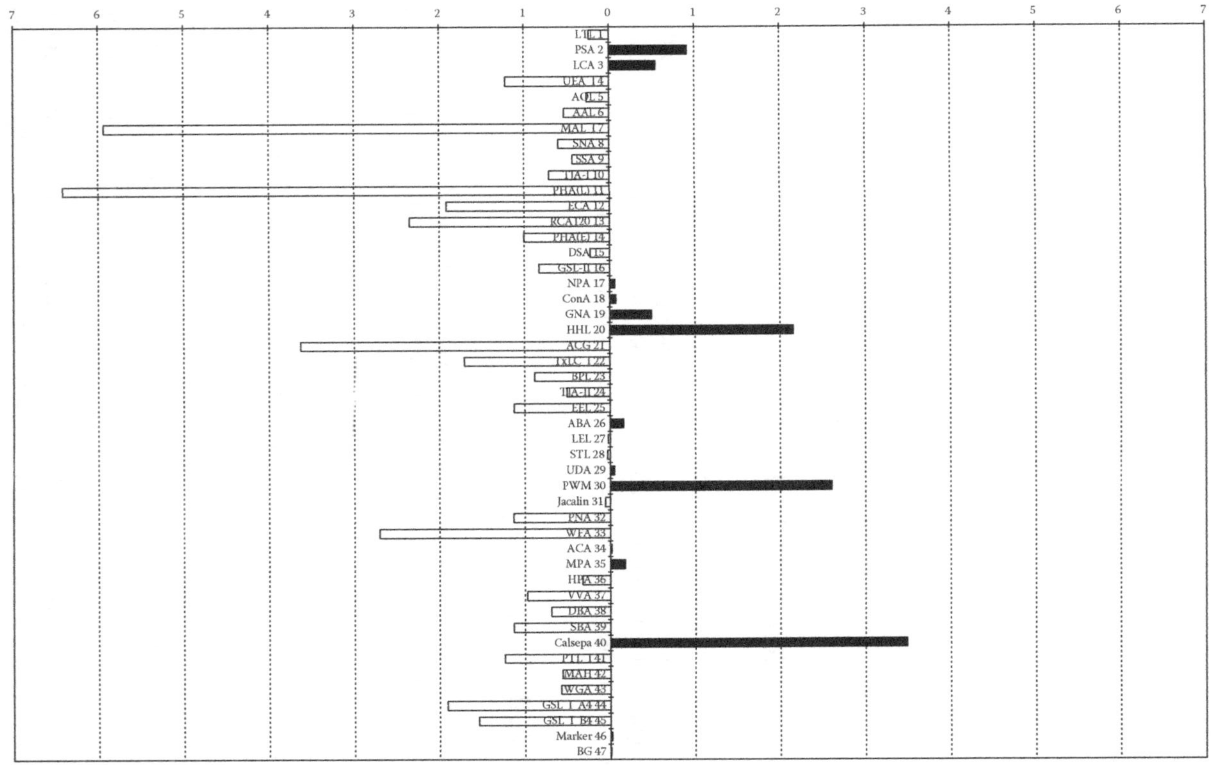

(b)

Figure 9.3
Differential profiling of Lec1 and CHO: (a) The signals from CHO are shown in white, and those of Lec1 are shown in black. (b) The rate of signal intensity change is more visible. The black-colored bars stretching to the right-hand side indicate increase in Lec1, and the white-colored bars stretching to the other side indicate increase in CHO. In this graph, “0” means no change between CHO and Lec1 levels.

4. Take out the LecChips (stored at −20°C), and place in an incubation box.
5. Apply 100 μL Probing Solution with an 8-channel multipipette to well of the LecChips.
6. Throw away Probing Solution, and place a LecChip face down on a paper* (e.g., Kimtex Quarterfold) to blot Probing Solution from the wells.
7. Apply 100 μL Probing Solution to the wells before drying up the wells.
8. Repeat steps 6 and 7 (Probing Solution washing three times total).
9. Throw away Probing Solution, and apply 100 μL of these samples to each well of the LecChips with a pipette.
10. Incubate the LecChips in an incubation box at 25°C for at least 16 h with shaking at a low speed.

Protocol 9.6: Scanning with a GlycoStation™ Reader 1200

Protocol 9.6.1: Reading the LecChip

Scan the LecChips with a GlycoStation reader, following the recommended reader settings and conditions. In order to detect very weak signals while avoiding saturation of strong signals, it is recommended that you take additional scans while adjusting the gain and the exposure time around the recommended condition.

Protocol 9.6.2: Data Analysis

Data is analyzed with the Array-Pro™ Analyzer Ver. 4.5 (Media Cybernetics, Inc., Bethesda, MD). The net intensity for each lectin spot is calculated by subtracting the background from the signal intensity.

Protocol 9.7: Example—Differential Profiling of Lec1 Mutants

Differential profiling is the most common methodology in assays using lectin microarrays. The result of differential profiling of Lec1 mutant cells (β1-2-*N*-acetylglucosaminyltransferase I-deficient mutant) and CHO cells is explained as an example of data interpretation.

Figures 9.3a and 9.3b show differential profiling of Lec1 and CHO.[2,3] Macroscopically speaking, the difference between CHO and Lec1 is more evident in *N*-glycan-related binders (e.g., PHA(L), PHA(E), RCA120, DSA, GNA, HHL, and Calsepa), whereas the difference in the signals of *O*-glycan-related binders

* We recommend a lint-free one like a nonwoven fabric.

(e.g., jacalin, PNA, ACA, MPA, HPA, VVA, DBA, SBA, and PTL-I) is relatively weak. The following are some examples of new discoveries that came to light easily as a result of differential profiling. (1) PHA(L) and PHA(E) are recognizers of branched complex-type *N*-glycans. It is quite reasonable that the signals of PHA(L) and PHA(E) dramatically decrease in Lec1, taking into consideration the characteristic lack of glycosyltransferase GlcNAc-T1; (2) it is also quite reasonable that the signals from high-mannose binders (GNA, HHL, and Calsepa) increase in Lec1; (3) the signals from lactose recognizers (RCA120) drastically decrease in Lec1; and (4) the signal from α2,3-sialic acid binder (MAL) decreases in Lec1, which is expected when there is a lack of glycosyltransferase.

It should be emphasized here that extensive fundamental data on lectin specificities are required for the interpretation of profiling data. For your convenience, some typical specificities of each lectin are summarized in Table 9.1.

Acknowledgments

The author thanks Drs. J. Hirabayashi, A. Kuno, and H. Tateno, AIST, for helpful advice and critical discussion about this work. The author also thanks all of the members of the Glycomics Research Laboratory, Moritex, Corporation.

References

1. Kuno, A. et al. (2005). Evanescent-field fluorescence-assisted lectin microarray: a new strategy for glycan profiling. *Nature Methods,* 2(11): 851.
2. Ebe, Y. et al. (2006). Application of lectin microarray to crude samples: differential glycan profiling of Lec mutants. *Journal of Biochemistry,* 139: 323.
3. Tateno, H. et al. (2007). A novel strategy for mammalian cell surface glycome profiling using lectin microarray. *Glycobiology,* 17(10): 1138.

Troubleshooting Guide

Note 1: Lectin List

This list is a collection from reported papers and Web sites. Moritex Corporation cannot provide a warranty for the appropriateness of these data for profiling data analyses, and assumes no liability for the results.

Note 2: Database

Moritex Corporation provides contract analysis services utilizing a comprehensive glycan–lectin affinity database (more than 10,000 interactions) that was developed at AIST. The database has now been made available to the public: http://riodb.ibase.aist.go.jp./rcmg/glycodb/LectinSerch. A differential profiling analysis software including an interpretation-assisting function is now available.

Note 3: Preparation of Cy3 tubes

3.1 Dissolve Cy3 by adding DMF 50 μL into a purchased Cy3 tube (usually, the tube contains the amount of Cy3 required for labeling of 1 mg proteins), and divide it into 10 tubes by 5 μL.
3.2 Dry up those Cy3 tubes by using "Centrifuge desiccator and Vacuum unit."
3.3 Put the Cy3 tubes in a shading bag containing desiccant agent of Cy3, and store it at 4°C.

Note 4: Spot Artifacts

If you see scratches or chipped patterns on lectin spots after the scanning, it is usually resulted from the dry-up process described in the paragraph 6 of Protocol 9.5 which means that the paper touched the surface of LecChips during blotting Probing Solution from the wells.

Chapter 10

Printing Methods

Todd Martinsky

Contents

Introduction

Microarray Technologies

Microarray technology over the last 12 years has become the analytical device of choice to unravel the molecular complexity of DNA, proteins, and other biomolecules at high speeds and resolution. Just as the microprocessor has revolutionized electronics and computer science, the microarray has dramatically transformed the way experimental biology is performed. A microarray is a set of microscopic spots of biomolecules on a planar surface, in an ordered array, which allows specific binding of other biomolecules in solution. In order to be considered miniaturized, the biomolecules spotted on the microarray must be smaller than a millimeter. If the array elements or spots in the array are larger than 1 mm, they should not be considered a microarray. The advantages to having microscopic spots on the array include high content, to analyze many biomolecules in parallel; fast hybridization kinetics; and low sample volumes. The miniaturization of the various microarray assays is the key to increasing throughput and reducing the cost. Planar substrates, such as glass, allow for automated manufacturing techniques to be implemented, low reaction volumes, and increased precision in detection. An ordered array of unique addresses for each biomolecule is required because we must know the correct location of every spot on the microarray to facilitate an accurate analysis of the data. An ordered array of rows and columns of spots allows for detection and analysis to be highly automated. Lastly, a microarray must have specific binding. Specific binding is required to accurately measure the number of molecules present in a sample. Each spot in the microarray should bind in predictable ways to its complementary partner in solution. Microarrays that do not allow specific binding cannot produce quantitative data.

Although a variety of technologies are used, all microarrays have the following:

1. Micro-size spots of immobilized biomolecules
2. One or more labeled samples that are hybridized to the microarray
3. A detection system that quantitates the hybridization signal

Just as the number of microprocessors on a single "microchip" continues to increase, so does the number of data points on the biochip known as the *microarray*. The increased throughput and lower cost of doing experiments in a parallel and miniaturized format have made microarrays a universally accepted biochemistry platform to meet a wide variety of applications. However, increased throughput is not the only advantage; reaction kinetics in a miniaturized biological assay are improved, allowing tests to be performed faster. Assay sensitivity is enhanced with miniaturization as the concentration of sample applied to a microarray-binding reaction is in dramatic excess to the very small spots that react with sample on the array. Additionally, smaller amounts of sample are required to perform tests. Microarrays reduce the amount of serum, blood, and other tissue types that must be taken from patients required to perform tests. This can be a critical aspect of pediatric testing when blood or serum is required, because there are strict rules on how much blood can be drawn from a child. Because of the microarray, in the not so distant future, all doctors will be able to perform as many tests as they would like, as quickly as they would wish to perform them, without worrying about drawing too much blood from a patient. For example, a single 100-spot protein microarray can replace 100 separate ELISA assays using 1/1000th the amount of sample. This is one small example of the types of problems that are being solved by microarrays. The advantages of microarrays have led to some diagnostic microarrays already receiving FDA approval.

Microarray manufacturing can be broken down into two main categories, *in situ* and *ex situ*. *In situ* manufactured microarrays are synthesized directly on the surface of the microarray base by base, which primarily makes them suitable for only DNA applications at this time. The main techniques of *in situ* manufacture are photolithography, ink jetting, and maskless lithography. In general, *in situ* techniques are not performed by the end user; this type of manufacturing is done by large companies. *In situ* manufacturing systems provided to end users do not allow any choices or optimization in the consumables used in the machines. Although *in situ* manufactured microarrays are limited to oligonucleotides, *ex situ* microarray manufacturing can put any premade material into a microarray format. DNA, proteins, carbohydrates, and other biomolecules are all being used. The two main types of *ex situ* manufacturing techniques are contact and noncontact printing. The dominating contact-printing technology for manufacturing microarrays is micro spotting pins. The dominating noncontact technology is piezo ink jetting. We will start by reviewing these microarray manufacturing techniques, and then examine the current scenario and future prospects of microscope-slide-size microarrays and examples of their utility. We will provide helpful information and protocols specifically related to manufacturing nucleic acid and protein microarrays in your own laboratory.

Photolithography makes use of semiconductor technologies. The process starts with a glass substrate containing a photomask, chemically prepared with specific locations to bind nucleotides. A photosensitive chemical that detaches under light from a mercury lamp caps each spot location on the microarray. The light activates modified photo-specific versions of the four DNA bases. The photolithographic masks ensure that the DNA synthesis is stimulated in defined positions and the masks predetermine which of the nucleotides are activated when flooded with one of the four types of nucleotides. The process of light activation and flooding the bases over the microarray is repeated until the bases are built.

Ink jetting can be used in both *in situ* and *ex situ* applications. Ink jetting for *in situ* synthesis is accomplished base by base in repetitive print layers using standard phosphoramidite chemistry. Inkjet heads, similar to those used on commercial inkjet printers, are connected to bottles containing the four different phosphoramidite nucleotides that make up the building blocks of *in situ* nucleic acid synthesis. Microarray manufacturing starts when the first nucleotide of each oligonucleotide is ink-jetted onto an activated glass surface of the microarray. Phosphoramidite synthesis reactions require the reactive sites on the nucleotides to be blocked with chemical groups that can be removed selectively to allow the bases to be added to each microarray spot, one base at a time, in a controlled manner. After the first base is ink-jetted, the nucleotide

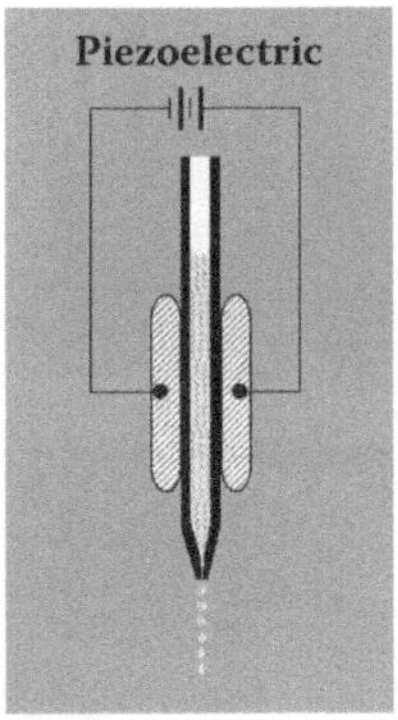

Figure 10.1
A schematic of a piezoelectric sample delivery mechanism.

is chemically prepared to react with the 3′ group on the next nucleotide. In between each step, the excess reagents are washed away so that they will not randomly react later in the synthesis. The process is repeated until all desired spots are formed. After the last base in the oligonucleotide sequence is ink-jetted, the microarrays undergo a final deprotection step to finish the oligonucleotide on the surface and make it available for hybridization. This technique has typically been used to make 60-mer oligonucleotides.

At the heart of the maskless lithography technique is a digital micromirror device that employs a solid-state array of miniature aluminum mirrors to pattern up to 786,000 individual pixels of light. The digital micromirror creates virtual masks that replace the physical chromium masks used in traditional photolithographic microarray manufacturing. A computer controls the desired pattern of UV light with individually addressable aluminum mirrors. The digital micromirror controls the pattern of UV light projected on glass in the reaction chamber, which is coupled to the DNA synthesizer. The UV light selectively cleaves a UV-labile protecting group at the precise location where the next nucleotide will be coupled. The patterns are coordinated with the DNA synthesis chemistry in a parallel, combinatorial manner such that up to 786,000 unique oligos can be synthesized in a single array.

Piezoelectricity is the generation of electricity or electric polarity in dielectric (nonconductive) crystals subjected to mechanical stress, or the generation of stress in such crystals when voltage is applied. A piezoelectric printing mechanism, as shown in Figure 10.1, uses a small dielectric crystal closely apposed to a fluid reservoir. Inkjet printers are based on this technology. In a typical configuration for microarray printing, a crystal is in contact with a glass capillary that holds the sample fluid. The sample is drawn up into the glass capillary, and voltage is applied to deform the crystal and squeeze the glass capillary to eject a small amount of fluid from the tip. Because of the ability to apply computer control and fast response time of the piezoelectric mechanism, this technology has been called *drop-on-demand*. The small deflection of the crystal results in drop volumes on the order of hundreds of picoliters to several microliters.

Ex situ micro spotting techniques work in conjunction with robots called *microarrayers* (Figure 10.2). The robot moves a spotting device, empowering the nanofluidic transfer of small spots of biomolecules to activated microscope-slide-size glass. One widely used device uses a combination of parts. One part is called *micro spotting pins,* and the other part, which holds the pins, is called the *printhead.* The key elements that make this pin technology work are twofold: a flat tip horizontally level to the centerline of the shaft of the pin and a defined exterior sample-loading channel. After being dipped into a solution of biomolecule mixed with buffer in a 384-well microtiter plate, the exterior sample channel fills with liquid and the flat surface at the end of the pin allows samples to be delivered to a substrate by capillary action. High mechanical tolerances of the pins are critical to consistent spotting. The printhead keeps the pins

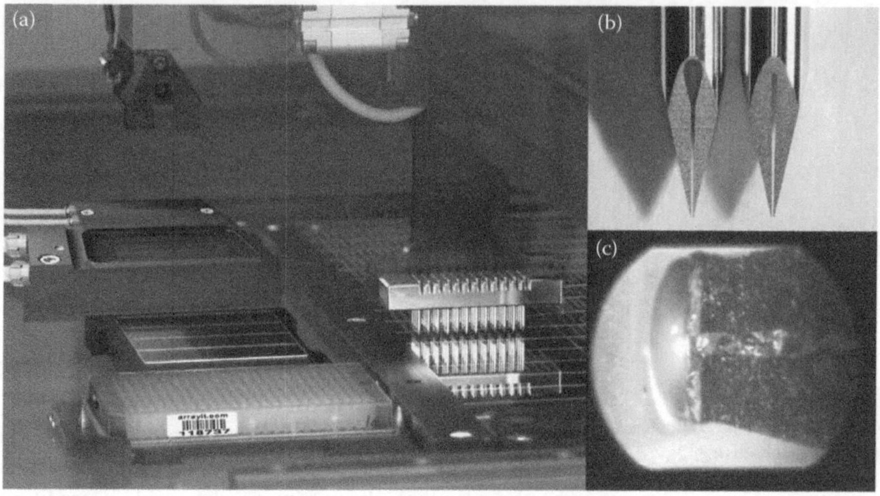

Figure 10.2
Inside a typical microarray robot: (a) A source plate, wash/dry station, substrates on the platen, and Stealth micro spotting device. The arrow is pointing to one tip of a 48 micro spotting pin tool. (b) Magnified images of the tips of micro spotting pins showing two different sample uptake channels with tips of identical size. (c) High-magnification photo of a sample being pulled off the horizontally level tip by surface tension.

straight and simultaneously allow them to individually float when they make gentle contact with the surface that they are spotting on. This keeps the force at the ends of the pins at a minimum, enabling them to last for many millions of consistent sample deliveries. The printhead also allows pins to be easily exchanged for different pins and patterns of pins, depending on the format of the arrays desired. The exterior sample channel makes them easier to clean between sample loading than closed capillary devices, thus making it easier to avoid cross-contamination of the biomolecules being spotted. The technology excels at printing multiple samples, multiple times over multiple surfaces with one low-volume loading of sample. Two goals are achieved at the same time, miniaturization on the microarray as well as miniaturization in the amount of sample used to make the microarray.

List of Protocols

Protocol Number	Name
10.1	Micro Spotting Pin Selection Performance Optimization
10.2	Surface Chemistry
10.3	Practical Considerations for Optimizing Protein Microarray Manufacturing
10.4	Processing Protein (Serum-Based) Microarrays
10.5	*Ex situ* DNA Microarray Manufacturing and Processing
10.6	The H25K Master Protocol

Protocol 10.1: Micro Spotting Pin Selection Performance Optimization

The desired diameter for the data-generating spots on the microarray is the primary decision-making factor in selecting a pin type to use on your microarrayer. Tips sizes for micro spotting pins range from 50 μm to 375 μm, making spot sizes ranging from 50 μm to over 500 μm. Often, the resolution of the detection instrument dictates the spot size required for a microarray. For reliable signal quantitation, 10 pixels of data within a spot are preferred, whatever the detection method. A 100 μm size spot detected at 10 μm resolution provides 10 pixels of data. Larger-tip pins make bigger spots and deliver more volume than smaller-tip pins. Because pins can easily be put in and taken out of the printhead by hand, it is possible to change spot size and delivery volume quickly. Ultimately, the spot size from any given pin size is determined by all the factors of capillary action. Surface hydrophobicity, sample viscosity, dwell time of the pin tip during liquid transfer, temperature, and humidity during printing all play their own roles in determining final spot size. Although there are a number of variables to control, it is not too difficult if you have the right tool set. In addition to tip size, there are four loading volume sizes from which to choose: 0.15, 0.25, 0.6, and 1.25 μL. The larger the loading volume of the pin, the higher the number of spots that can be printed with a single dip into a source plate. Pins are most commonly held in the printhead in up to a 4 × 12 pattern at 4.5 mm center-to-center distances. When many samples are printed, all 48 pins are used, whereas for small low-density microarrays, perhaps only a single pin is used.

Getting a set of micro spotting pins to work well for any given sample type and surface chemistry combination requires optimization. Speed settings of the microarrayers axis's, wash, dry parameters between sample pickups, dwell times of the pins to the surface being printed during sample delivery, tip size, buffer composition, surface chemistry, and environmental conditions all must work together for best spot morphology and consistent spotting. Most microarrays are manufactured on activated glass surfaces, but some are made on 3D surfaces such as hydrogels, nylon, nitrocellulose, and polyvinylidene fluoride (PVDF) membranes. When printing on a fragile membrane surface, it may be necessary to use slower *Z*-axis speeds to keep from denting the membrane during the manufacturing process. Nylon membranes that bind DNA are typically hydrophilic, so keeping the pin dwell time on these surfaces at a minimum to avoid depositing too much sample on each touch off is critical. The software of the microarrayer controls the number of spots from each loading of sample from the source plates; in general, when loading the same sample, pins are not cleaned and dried between loadings. How far a pin travels in the *Z*-axis toward the substrate slide after the tip has made initial contact and how quickly it approaches and moves away determines dwell time. These movements are controlled by the microarrayer robotics; therefore, without accurate and repeatable *X*, *Y*, and *Z* movements from the robotics, it is impossible to manufacture good microarrays. The best microarrayers now measure submicron movements with encoders to make sure that the proper movements are executed, to guarantee that all movements in a given program are highly repeatable.

Protocol 10.2: Surface Chemistry

It is important to understand the principles that determine how microarray surface chemistries and spotting buffers are used and interpreted before undertaking any new microarray manufacturing project. Microarray surface chemistry must attach biomolecules in a stable manner. A stable attachment that can handle all the processing steps the microarray will go through is essential to producing strong signals and accurate results. Capture molecules must be immobilized, remain in a specific location, and stay essentially unchanged throughout all the pre- and postprinting procedures of the microarray experimental life cycle. Only a stable surface can meet this need. However, the stability of the surface cannot be at the expense of maintaining the activity of the biomolecule. The binding, hybridization, epitope, or enzymatic activity properties of biomolecules are essential to usability of the microarrays. Without the binding activity of the printed sample, there can be no experiment. Most DNA and many protein sample types are quite stable when they are clean and dry after spotting, so covalent, electrostatic, and simple absorption approaches to immobilization have proved to work effectively. Activity of less stable molecules or molecules that may lose their function if they denature, such as some types of proteins, enzymes, and others, can be prolonged on a surface by using stabilizing printing buffers or specialized 3D surfaces to preserve the spotted sample. Finding the optimal buffer that maintains the activity of a biomolecule on a surface that is both compatible with the surface chemistry and provides good spot morphology can take some work to accomplish. Commercial manufacturers of microarray surfaces often have specific buffers formulated or suggestions on what to use. In some cases, the biomolecule should be allowed to dry on the surface, and in other cases, it is desired that the spots printed stay wet. For example, proteins subject to denaturing when dry can be spotted in 40% glycerol on epoxide surface chemistry and stored at 4°C. Oligonucleotides being attached to amino-silane glass slides by electrostatic charge interaction and subsequently UV-cross-linked must be dry to allow the DNA to cross-link to the surface.

Microarray manufacturing takes time; in some high-density spotting runs, a set of microarrays may take days to finish. It requires that the surface used maintain its binding activity and remain unchanged throughout the microarray manufacturing process. The environmental conditions of the microarray chamber have to be compatible with the surface. Additionally, time and flexibility are required to design experiments, plan manufacturing, procure product, prepare samples, and store printed microarrays prior to using them. Knowledge of the expiration date of the surface must be known as well as the shelf life and proper storage conditions for a microarray after manufacturing.

To generate quantitative data, the density of reactive groups and/or attachment method on the surface must be uniform across the entire spotting area of the substrate to ensure identical coupling of biomolecules at each microarray location. The molecules attached to the surface must reflect their specific binding characteristics accurately. Samples must not be allowed to attach to the source plate and reach the surface chemistry at the proper concentration. Nonbinding plates for proteins and polypropylene plates for DNA samples are routinely implemented. Polystyrene plates are not desirable as source plates as they are positively charged, and can bind both DNA and proteins. Lateral flow of the sample on the surface must be kept at a minimum to keep spots defined and the density of bound molecules within a spot to optimize reaction kinetics. The concentration of sample in relation to reactive groups on the surface needs optimization as well. Without homogeneity, different amounts of biomolecule will attach in different areas of the substrate and compromise quantitation. Regardless of the printing technology (contact or noncontact), the samples being printed at a spot location are being deposited in saturating volumes to the attachment sites of each location. What does not attach to the surface in each spot is washed away in the preprocessing steps of a microarray. Spotting of too high a concentration can put too much unbound biomolecule on the microarray, making it difficult to wash and causing unwanted background noise; too low a concentration, and the microarray may not be able to generate a proper signal.

Maintaining surface flatness and parallelism is important to achieve high-quality printing and detection. Accurate robotics and sophisticated printing devices are used to deposit samples. That means the distance between the printing surfaces and printing mechanism must be calibrated within a minimum of 100 μm for high-quality printing to take place. Unlike contacting printing, spot size of noncontact delivery systems is determined by how far away the delivery nozzle is from the printing surface. In other words, the farther a sample travels through the air, the bigger the spot gets. Surfaces that deviate in height will compromise spot quality. Additionally, current microarray scanners have focal planes in the resolution range of 5–50 μm. Accurate planarity of the substrate is required to maintain proper focus for acquisition. This is not a problem if your detection instrument has dynamic autofocus, but many microarray detection instruments do not have this feature. Reacted microarrays must stay in focus during scanning, and the physical properties of said substrate and reacted immobilized spots should not change during or after scanning.

Fluorescent detection instruments dominate the microarray industry; however, colorimetric, chemiluminescent, light scattering, surface plasmon resonance, planar waveguides, and others are being implemented. Regardless of the detector, the surface must not significantly contribute to the background noise of the detection system. For example, surfaces such as nonreflective, clean, white membranes read as zero background on some colorimetric systems and flatbed scanners. Transparent glass is the preferred material for fluorescent-based detection devices. Nontransparent surfaces reflect photons into the PMT or charge-coupled device (CCD) detector and this leads to higher backgrounds from back scattering of light.

The same physical size of each substrate facilitates automation for manufacturing and processing, dramatically increasing microarray usability. The standard in the microarray industry is 25 mm × 76 mm × ~1 mm. ArrayIt, Agilent, Roche/NimbleGen, and many other organizations have standardized on the microscope-slide-size glass format, but microarrays in the bottoms of standard microtiter plates are becoming more common. Using standardized formats ensures compatibility with open microarray platforms, reduces cost, promotes innovation, and increases flexibility for the end user by empowering the use of products from multiple vendors.

Each substrate must be the same to guarantee accurate results, and some organizations polish the glass for more homogenous topography and, thus, more consistent results. A surface used in an experiment must work in the same way from one day to the next, and one experiment to the next. Without consistency, it is impossible to plan, design, execute, analyze, and compare experiments over time. A laboratory running a microarray test could use dozens to hundreds of microarrays per day. That means only one laboratory and one application could easily use 25,000 substrates per year. Any surface must meet all the rules presented here and allow for mass production rates to guarantee availability. Surfaces prior to and after sample immobilization that do not require special storage conditions and have a long shelf life will be most desired. The ability for mass production allows for economies of scale, which leads to the most important characteristic, *product affordability.*

Advantages of 2D surface chemistry for microarray (e.g., activated homogeneous glass) are the following:

1. Inherent lower background in fluorescent detection (glass)
2. Better-defined spot morphology (no diffusion)
3. Compatible with SPR, planar wave guides, RLS, and other exotic detection strategies
4. High specificity of binding
5. Nonporous surface (no place to trap any contaminate in processing)
6. Covalent and/or specific binding to avoid altering biological activity

Advantages of 3D surface chemistry for microarray (e.g., membranes, filters, and gels) are the following:

1. High binding capacity (absorption)
2. Compatible with fluorescent, chemiluminescent, and colorimetric detection
3. Longer history of use (comfort level for users familiar with nitrocellulose, nylon, and PVDF)
4. Less expensive labeling reagents and reading equipment (colorimetric)
5. 3D capture moiety to avoid altering biological activity

Protocol 10.3: Practical Considerations for Optimizing Protein Microarray Manufacturing

Protocol 10.3.1: Experimental Design

What peptides and/or proteins are important in the system of study is often the most difficult question to answer in any microarray experiment. DNA microarrays are much easier because the full genomes of most model organisms are readily available. Protein microarrays often rely on existing scientific literature as the starting place to build a list of potential array element proteins or peptides. Users may also wish to add content based on proteins derived from gene expression experiments and other educated decisions based on potential mechanisms for their biological function of interest to test out a new hypothesis. Additionally, every microarray must have positive and negative controls. Positive controls to validate the microarray are important to confirm the findings of other groups. Controls often include marker spots, which are always present and provide a means for easy spot finding and downstream data analysis. The goal of microarray experimentation is to investigate a high number of proteins simultaneously, not to get every array element recognized. Therefore, the goal of microarray design is to create a microarray that will allow experiments to generate data that will allow a predictable profile to emerge, which can be used to extract meaningful information.

Protocol 10.3.2: Selecting Peptides/Proteins

Proteins for microarrays come from a variety of sources, and all types of proteins are put into the microarray format as binding targets. Proteins are often isolated from tissue. The benefit of this method is that native proteins most likely are already known antigens. The big downside, however, is that the method is time intensive, thus making it difficult to build a large list of array elements. Proteins, of course, can be purchased from a variety of commercial sources. However, this can be expensive when one wants hundreds, or perhaps even thousands, of proteins for their microarray. Recombinant whole proteins do not provide specific epitope-binding information. Peptide synthesis of custom peptides allows for excellent flexibility and is coming down in price, with most large vendors providing discounts on bulk synthesis of many peptides. Regardless of the source, microarray manufacturing requires only small amounts of protein, so luckily, a little bit of sample will go a long way.

Protocol 10.3.3: Sample Preparation

Purchased peptides or proteins are most often shipped lyophilized and must be made soluble in a printing buffer that will make the protein soluble, stabilize the sample, provide efficient transfer onto the surface chemistry, and optimize the attachment to the microarray substrate. Antibodies and antigens, in general, are much easier to handle than peptides. Getting peptides in solution can be hard work; the complexity of the surface chemistry type can impose limitations on what buffers can be used. Some hydrophobic peptides are not soluble in aqueous-based buffers, so DMSO is often used. High concentrations of DMSO dissolve nitrocellulose; therefore, this buffer should not be used with nitrocellulose microarray surfaces. The overall charge is a major factor of solubility, and one can calculate the net charge to determine whether the peptide is acidic, basic, or neutral. From this information, one is guided to select different solutions as good starting points for getting peptides to go into solution. Thus, there are really two problems with reconstituting peptides. The first is getting the peptide into solution, and the second is having a solution compatible with the surface chemistry such that the spotted peptides will attach. In addition to solubility, concentration is also important. All microarray surfaces have a fixed concentration of biomolecule that will bind to the surface, and what does not attach will wash away in processing the microarray, which is done just prior to the binding reaction. Most microarray surfaces cannot bind a concentration higher than 1 mg/mL, and anything less than 0.1 mg/mL as a final concentration is typically not enough to generate a good signal. The goal (if one is not printing a dilution series) is to spot a concentration that ensures the binding sites of the surface chemistry are saturated and keep the excess to a minimum. Excessive amounts of unbound protein can be difficult to wash away and cause undesirable noise in the experiment. In my laboratory, we have developed two fundamental approaches for peptide microarrays, one using biotinylated peptides on our streptavidin surface chemistry and the other amino-linked standard peptides spotted out on our epoxide chemistry. Biotinylated peptides are made soluble with the least amount of DMSO possible, as anything over 20% will make the sample incompatible with the streptavidin surface chemistry. Epoxide chemistry is compatible with any percentage of DMSO, but in general, a maximum of 50% DMSO is used.

Protocol 10.3.4: Setting up the Source Plates for Printing

As discussed, always use positive and negative controls. It is a good idea to use marker spots that are sure to be reactive for the positive controls. Spotting buffer is only a good choice for the negative controls. For best results, controls should be uniformly distributed throughout the slide. This will allow for easy quality control of the uniformity of the surface and guarantee that if there is a bad area on the surface, it will not ruin the entire experiment. All manufacturing processes have some variability, and therefore consider the number of replicates required to ensure that good statistics can be applied to the raw data to determine an average value that reflects an accurate value for molecules binding to the spots of unique proteins on the microarray. In general, when pin spotting, 1 to 4 replicates is considered sufficient; anything higher has proved not to be statistically useful unless the microarray quality is very poor. The number of times a source plate can be reused depends on the stability of the samples in the source plates. In general, synthetic peptides are quite stable, and the source plates can be reused several times. Dehydration is a factor, so plates must be monitored so that sufficient volume exists for loading samples to the spotting pins. Sufficient volume depends greatly on the geometry of the plates; in general, 384-well plates with round wells and conical bottoms provide the best source plate type. This geometry coupled with micro spotting pins can load sample from volumes as low as 2.5 μL. Avoid polystyrene plates as they are intrinsically positively charged and will bind proteins and peptides; additionally, spot-ready samples in these types of plates form a meniscus, thus increasing the volumes in the plates required to get a good loading of sample. Once plates are set up to maximize stability, be sure to avoid as many freeze–thaw cycles as possible. As the source plates are so important, it is prudent to make a few replicates and store them for future use.

Protocol 10.3.5: Executing the Print Run

How It Works

When setting up the plate formatting, the number of micro spotting pins needs to be considered. The rule is, the lower the number of array elements, the lower the number of pins used, and vice versa. Multiple microarrays can be spotted on a single slide substrate, so any complete set of spotted array elements can be considered a microarray. Because multiple pins are used to make microarrays, what a single pin prints from a given sample set can be considered a subarray. Subarrays are also often referred to as *subgrids* and *blocks*, and are defined as the set of samples on the microarray spotted from a single pin. Therefore, a microarray printed with 4 pins has 4 subarrays, and a microarray printed with 48 pins has 48 subarrays. Smart microarrayers allow for the programming and sample tracking of any pin configuration and sampling pattern as long as the pin tool and plate are compatible. For example, it is possible to use a pin tool with 9 mm center-to-center spacing of pins and load samples from a 384-well plate, but it is not possible to load a pin tool using 4.5 mm centers from a 96-well plate. Working through the logic of the pickup pattern and where spots are going to end up on the microarray, carefully consider and use good software to track where each sample is located on the microarray.

Required Materials

1. NanoPrint microarrayer equipped with 946MP3 micro spotting pins
2. Spot-ready samples in 384-well plates
3. SuperEpoxy 2 microarray substrates

Step-by-Step Protocol

Step 1: Set Up the Pin Type Through empirical studies of printing test samples with a specific pin type, the pin parameters are determined and programmed into software (Figure 10.3). Tip size and uptake volume are fixed on the basis of the physical geometry of the pins. In this example, the pins are programmed to reload after every 150 spots printed, not wash prior to reloading, and spots cannot be programmed to be closer than 120 μm center-to-center distance in the print design section of the software.

Step 2: Set Up Pin Tool This example shows the maximum number of pins that are typically used on microarrayers. The reason this 48-pin configuration is most popular is that it maximizes the number of spots that can be printed onto a 2 mm × 75 mm × 1 mm slide substrate, while still maintaining an area for a barcode or other label at the bottom of the slide for identification. The printable area is also compatible with the working area of most automated slide-based hybridization systems and manual processing methods (Figure 10.4).

Step 3: Create Sample-Loading Sequence for the Print Run A full complement of 48 pins will use eight loading locations for a single 384-well plate. Each plate can have a unique loading sequence if desired, allowing samples to be picked up and printed in any order (Figure 10.5). In general, it is not important for spots to be in any specific location on a microarray. What is important is that the location of spots

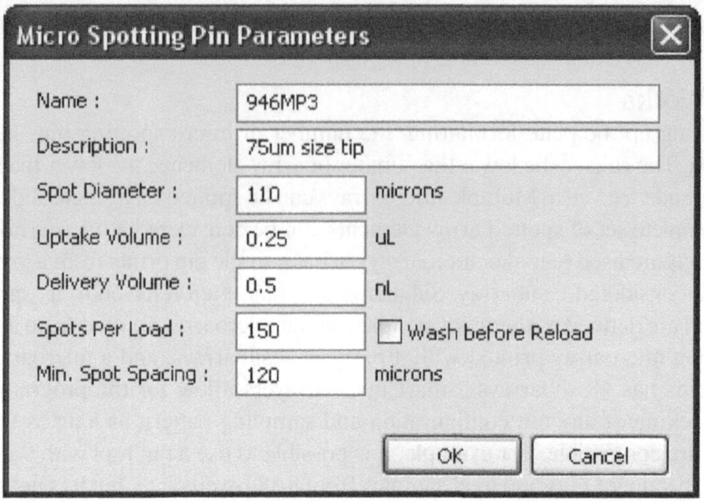

Figure 10.3
Micro spotting pin parameters.

be known and tracked. Full plates are easily set up by selecting all locations for full plates; plates that are not full are programmed by clicking the specific plate location. Loading patterns are saved individually and used later on in programming to define a plate list for the print run (Figure 10.6).

Step 4: Setting up the Microarray Print Design There are four tabs in the print design dialog box (Figure 10.7). The first defines the pin configuration and type,

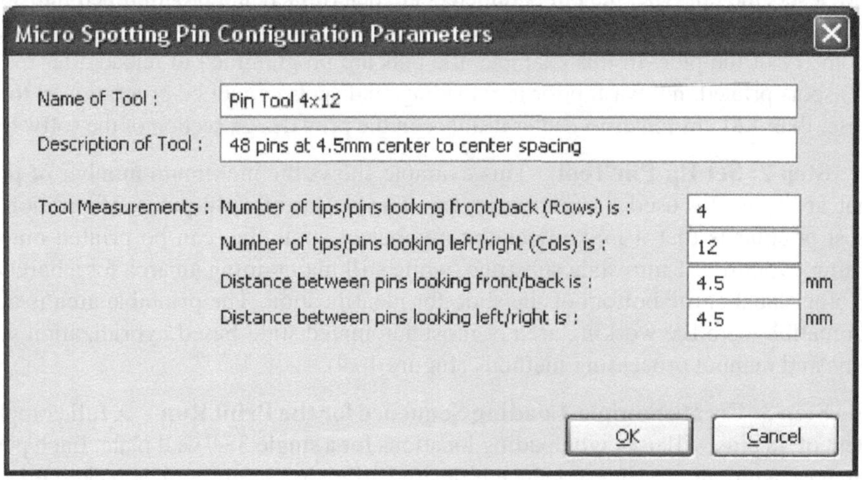

Figure 10.4
Pin configuration parameters.

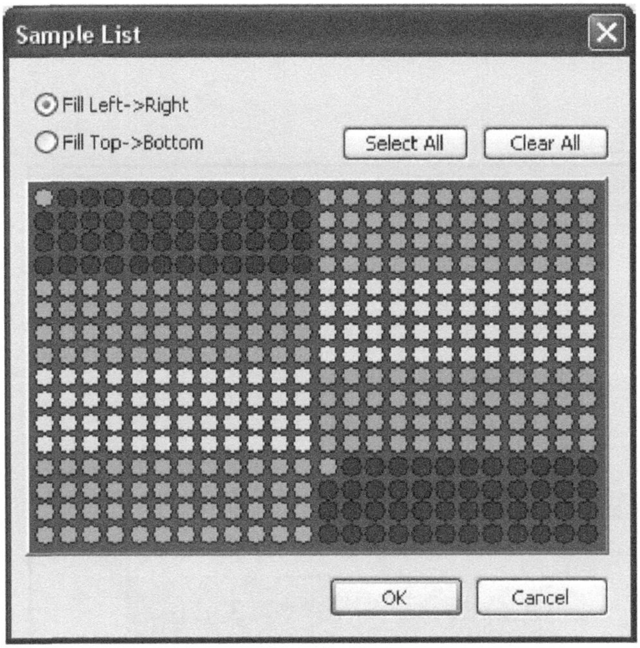

Figure 10.5
Sample sequence.

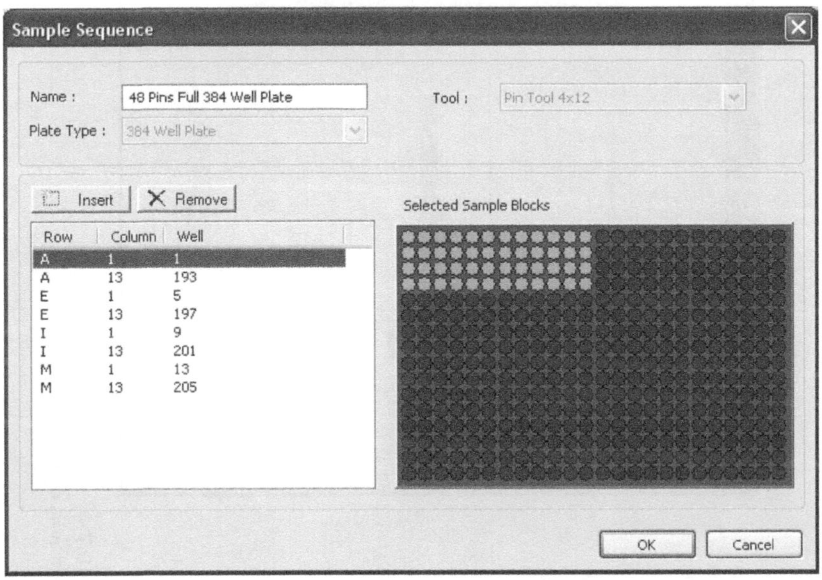

Figure 10.6
Sample list.

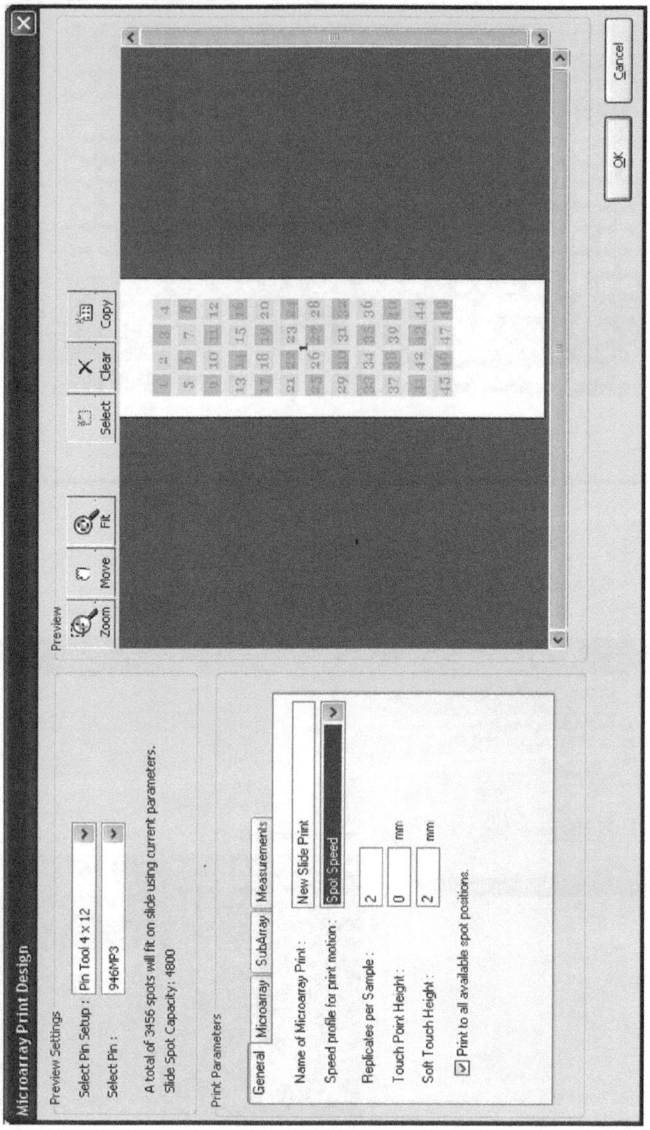

Figure 10.7
Microarray print design.

name of the print design, speed settings, number of replicates for each array element, touch point height, soft touch height, and whether all spot positions should be printed to. Touch point height allows the end user to make adjustments to how far pins move in excess up or down from the main calibration location of the Z-axis movement as it relates to each slide or substrate on the work table of the microarrayer. The speed settings in software allow for the control of two separate movements in the Z-axis: the gross motor movement and the soft touch. Soft touch is the last 2 mm of the Z-axis movement and is a different speed from the gross Z-axis motor movement. This is done to minimize wear on pins as well as to maximize print speeds.

The "Microarray" tab defines the number of microarrays to be printed on the slide substrates and their spacing (Figure 10.8). When using 48 pins, it is not possible to get more than one microarray on a slide. However, this feature in software is very useful for spotting "arrays of arrays" on single-slide substrates, or when spotting microarrays to the bottoms of 96- and 384-well plates.

The "Subarray" tab defines the subarray dimensions (Figure 10.9). Each subarray is numbered 1–48. As the set of pins all move together, each subarray for any given pin configuration has the same number of columns and rows. Depending on the spacing set in the pin setup, the software provides some guidance in the maximum number of columns and rows for each subarray. Using the check boxes, the direction the pin set moves over the slide substrates can be changed in addition to increasing the density. Double-density printing uses an "orange pack" or interstitial print pattern to allow spots to be printed at closer center-to-center distances.

The fourth and final tab is the "Measurements" tab, which is used to set the location of the microarray on the slide substrate (Figure 10.10). This tab also defines the slide size on the work table. Work tables can be made to accommodate any size slide, substrate, or plate.

Step 6: Set up the Sample Plate List Our print design uses 48 pins and duplicate spotting, each pin, making a 36 × 36 subarray for a total of 62,208 spots. Therefore, 81 384-well plates full of samples will be needed (Figure 10.11). If the microarrayer is equipped with an autoloader, these source plates can be fed automatically by the autoloader. An Excel spreadsheet of all source plates must be created for sample tracking (Figure 10.12).

The user puts unique ID, name, and comment for each well of each plate. Plate number, row, and column information remain unchanged. After all parameters for the print are created, this file is run through the program to create the spot map (Figure 10.13). Spot maps used by microarray scanners to locate spots on the microarray will be covered later in this chapter.

Each plate loaded into a plate list is associated with a sonicate–wash–dry protocol. In general, the same wash protocol for pins between each different sample pick is the same for all source plates.

Step 7: Final Method Properties Final method properties is a summary that includes the number of slides and other parameters needed for execution of the print run (Figure 10.14). After printing, spots are visible on the slides and can be checked for quality using a standard stereo microscope.

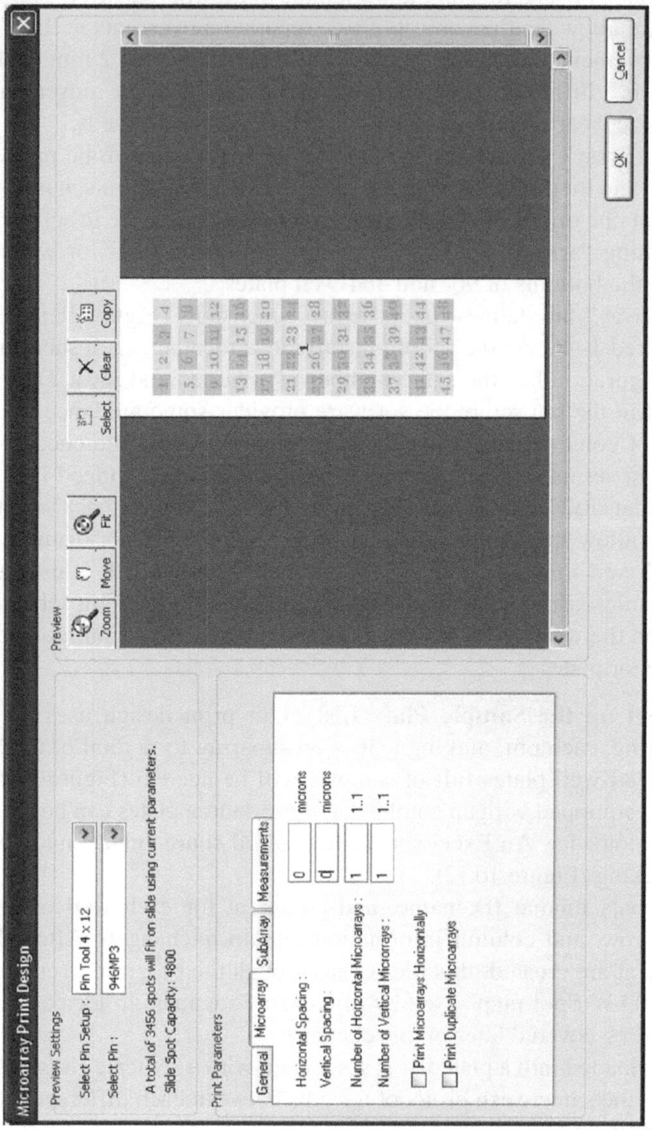

Figure 10.8
Print design: Microarray tab.

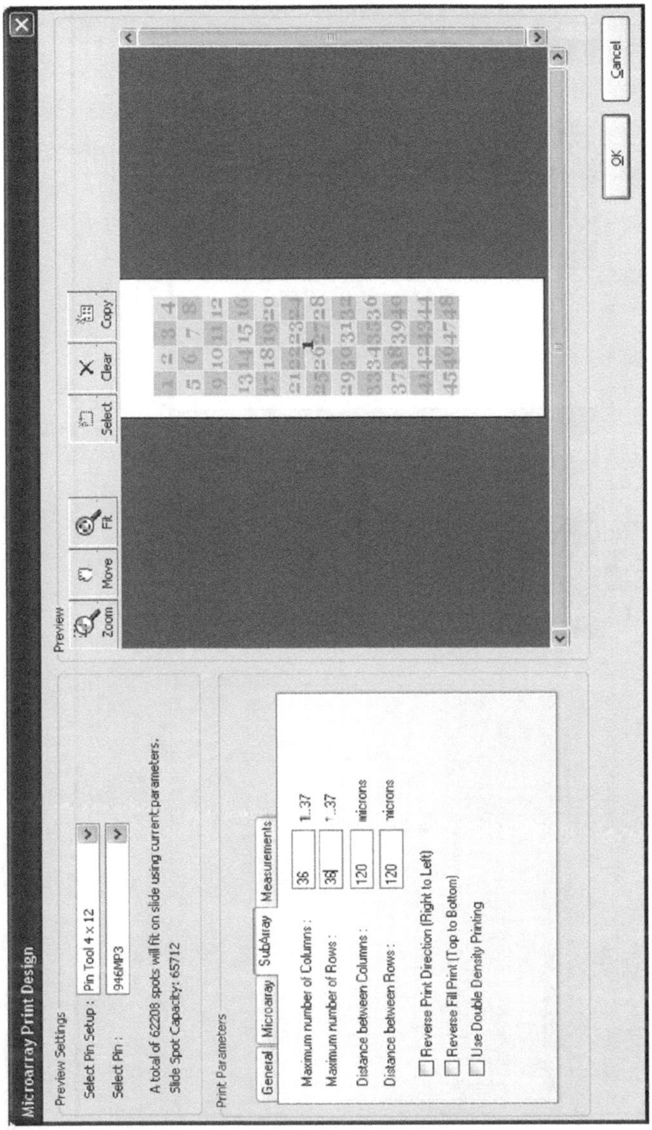

Figure 10.9
Print design: Subarray tab.

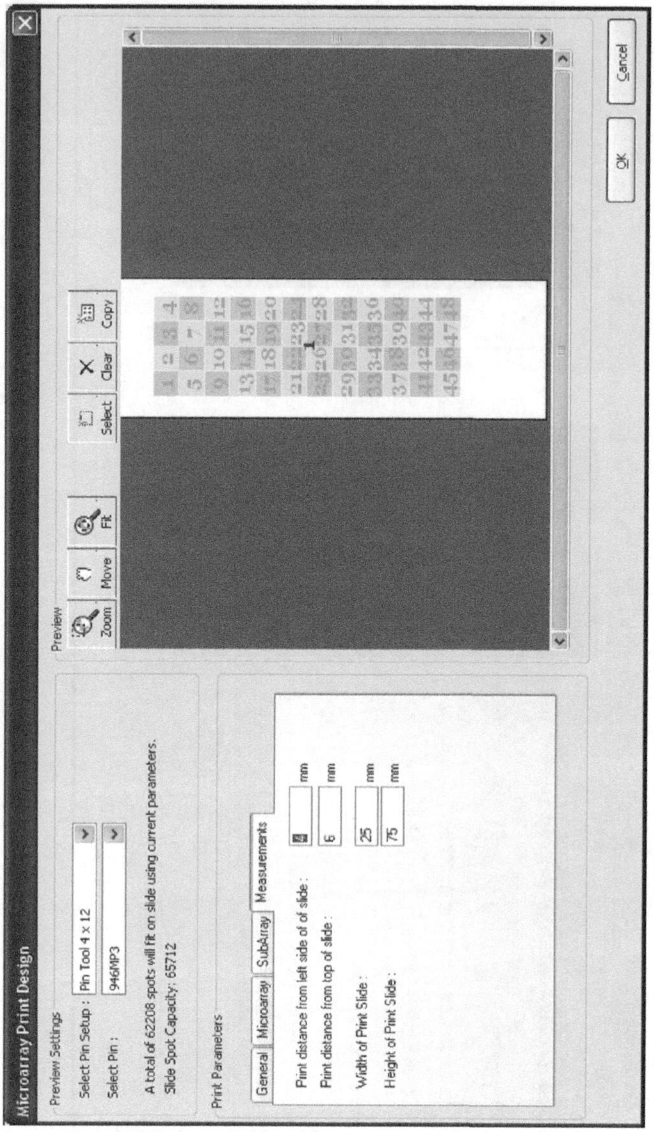

Figure 10.10
Print design: Measurements tab.

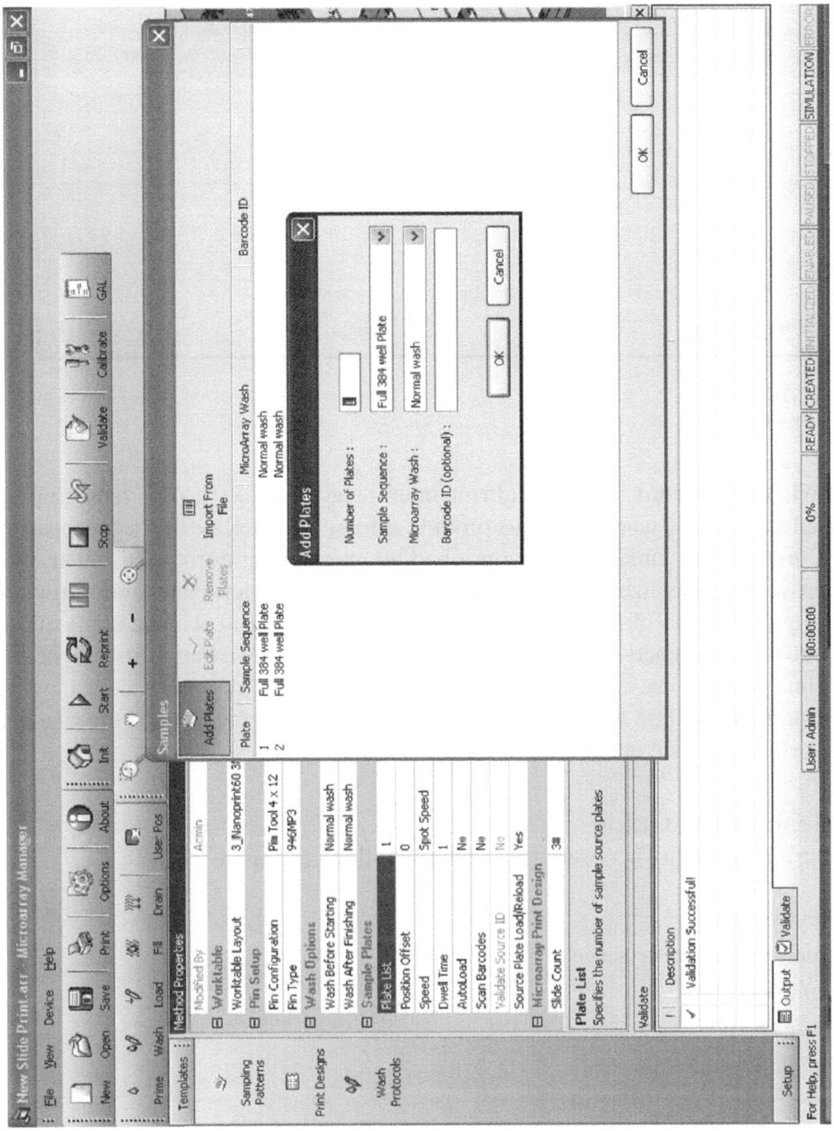

Figure 10.11
Sample plate list.

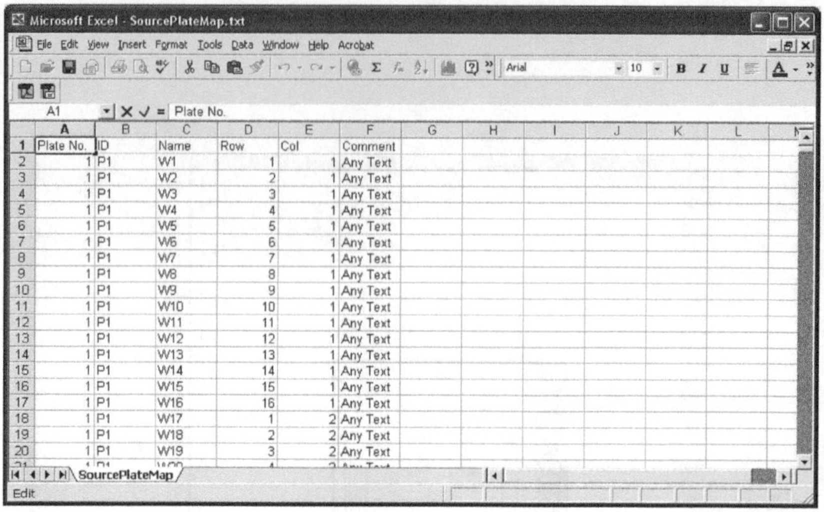

Plate No.	ID	Name	Row	Col	Comment
1	P1	W1	1	1	Any Text
1	P1	W2	2	1	Any Text
1	P1	W3	3	1	Any Text
1	P1	W4	4	1	Any Text
1	P1	W5	5	1	Any Text
1	P1	W6	6	1	Any Text
1	P1	W7	7	1	Any Text
1	P1	W8	8	1	Any Text
1	P1	W9	9	1	Any Text
1	P1	W10	10	1	Any Text
1	P1	W11	11	1	Any Text
1	P1	W12	12	1	Any Text
1	P1	W13	13	1	Any Text
1	P1	W14	14	1	Any Text
1	P1	W15	15	1	Any Text
1	P1	W16	16	1	Any Text
1	P1	W17	1	2	Any Text
1	P1	W18	2	2	Any Text
1	P1	W19	3	2	Any Text

Figure 10.12
Source plate map.

Work table layout. Our NanoPrint microarrayer allows a user to set up an unlimited number of work tables to provide ultimate flexibility to the microarray robot. Current work tables that are designed and running at current customer sites are microscope-slide-size glass, microfluidic cartridges, microplates, and proplate glass. Any other type of sensors used in biology and material science applications can be custom-designed and implemented quickly. The existing software of the NanoPrint is easy to use and flexible enough to print various targets. Once you own a NanoPrint, there will be no need to purchase a new microarray robot if the size and shape of your substrate changes.

Troubleshooting Guide

If poor printing quality is observed, check the following 5 areas:

1. Pin Condition
2. Microarrayer Performance
3. Sample Quality
4. Slide Quality
5. Environmental Controls

1. Pin Condition Inspect pins for physical damage under a low power objective (e.g. dissecting) microscope. If they are physically damaged, the tips are too small to repair and need to be replaced. See Figure 10.2(b) as well as Appendix Figure A11 for examples of undamaged and damaged pins. However, if pins do not appear to be damaged but are not loading, sample cleaning is recommended. The

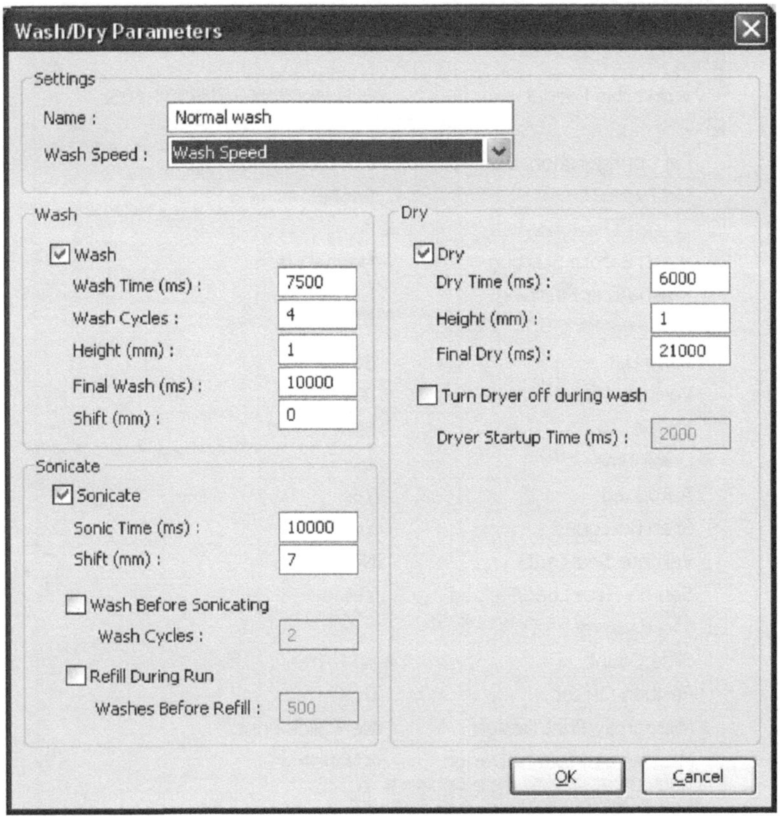

Figure 10.13
Wash or dry parameters.

following cleaning procedure will condition your Micro Spotting Pins to print virtually any type of sample. It is also recommended to use this cleaning procedure after every print run.

Cleaning Procedure Wear powder free nitrile gloves throughout this entire process. Touching pins with latex gloves can leave contaminating residue on the shafts of the pins.

Step 1. Sonicate pins for 5 minutes in a 2-5% concentration of Micro Cleaning Solution (ArrayIt catalog # MCS). Suspend pins in the sonication bath such that they do not touch the sides or bottom of the bath. Longer sonication times may be required if material has dried on the pin.

Step 2. Rinse pins under a stream of hot tap water for 2 minutes or longer to assure that detergent is completely rinsed away.

Step 3. Sonicate pins for 3 minutes in steam-distilled water. Never expose pins to de-ionized water. De-ionized water pulls ions out of the stainless steel of pins thus promoting corrosion.

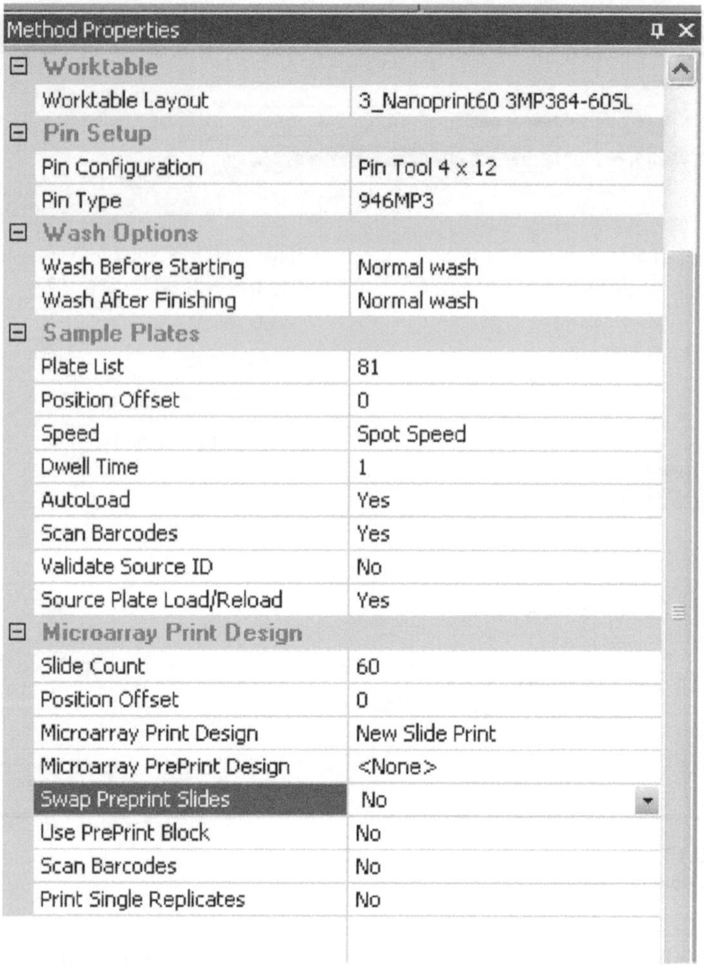

Figure 10.14
Method properties.

Step 4. Carefully blow dry with clean forced air. Do not use air from canisters. The canister propellants will destroy the surface tension properties of the pins. Never load wet pins into the print-head. This will cause the pins to stick during printing and can also lead to corrosion.

Step 5. Place a protective cap over the tip of each pin. Cleaned pins should be stored in a clean, dry, closed container. Do not handle pins without gloves.

Verification You can easily determine if your pins are printing by observing the spotting of the solution while the pins are arraying your samples. Good commercial print buffers leave visible spots during the arraying process. They are visible due to the salts in the solution and will disappear during processing of the microarrays leaving an invisible array of your attached biomolecule on the microarray substrate. After printing you can hold the array up to a light source, or view under a microscope and usually visually determine if the spots were printed and if there are any gross defects such as spot merging or missing spots. Finer evaluation of the array may require a test hybridization and microarray scan.

2. Microarrayer Performance Verify that pins leaving the microarrayer wash dry station are completely clean and dry in order to load a new set of samples. If they are not, they will not load completely, resulting in a reduction in the number of spots printed as well as spot quality.

Check that the microarrayers movements are repeatable. Most microarrayers utilize an encoder (e.g. NanoPrint, Arraylt), which can trace every movement the machines make and generate a .txt tile which can then be reviewed. Arraylt recommends not less that 10-micron repeatability/accuracy in all three motions x, y and z.

3. Sample Preparation Samples for printing should be consistent in concentration and viscosity. Well-to-well variations will lead to irregular printings amongst pins. Do not fill the source plate (sample) wells with excessive volume. Pins pick up samples by capillary action from the end of the pin tip and not its shaft. Filling the sample wells at heights much greater than I mm can cause large spots to form. This is caused by too much sample sticking to the outside of the shaft which drains down onto the spotting surface.

4. Surface Chemistry Using standard microscope slide glass as a printing substrate is not recommended. By nature, standard microscope slide glass is not homogenous. Instead, use optical quality glass processed in a cleanroom. It is also important to have a flat printing surface for consistent delivery of sample. Standard microscope slide glass does not have consistent physical dimensions for this purpose. Difference in thickness changes the dwell time of micro spotting pins on the surface. The longer the pin remains on the surface, the more sample gets delivered. This makes it more difficult to get a consistent delivery of sample.

5. Environment Verify that the slides have been stored properly in a closed container and are free of debris. Control measures for cleanliness, temperature and humidity must be in place. Dust is the microarrayers worst enemy. Keep all areas of the microarrayer as clean as possible. Using microarrayers in cleanrooms helps a great deal but is not necessary to print good microarrays. A "cleaner" room can be achieved using commercial HEPA filters and common sense. Always wear gloves. In general keep the working environment dry and comfortable; add humidity to the microarraying chamber. Microarray humidity should be set between 45-65% RH and regulated to ± 2% RH. Lower humidity settings will cause printed spots so be of smaller diameter.

Microarray Terms Used

Pin Setup

Pin Configuration: The pin configuration to use when printing.

Pin Type: The type of pin to use when printing.

Wash Options

Wash Before Starting: Wash pins before starting print method.
Wash After Starting: Wash pins after finishing print method.

Print Design

Slide Count: The number of slides to print.
Position Offset: The number of slide positions to offset the first print slide. The remaining print slides must be placed on the slide rack sequentially.
Microarray Design: The type of Microarray Print Design to use when printing. See Microarray Print Design.
Microarray PrePrint Design: The type of Microarray PrePrint Design to use when preprinting.
Use PrePrint Block: Specifies if a preprint block is being used instead of preprint slides.
Scan Barcodes: Specifies if barcode identification is being used for the print slides.
Print Single Replicates: Prints each replicate to each slide before printing the next replicate.

Sample

Plate Count: The number of sample source plates.
Position Offset: The number of source plate positions to offset the first source plate.
Speed: The speed profile to use when printing.
Dwell Time: The time to pause when the microarray pin is loading.
AutoLoad: Specifies if an external plate loading device is being used.
Scan Barcodes: Specifies if barcode identification for the source plates is being used.
Validate Source ID: Specifies if the external loading device should validate the barcode ID of the source plate with the user-assigned ID of the sample source plate.
Source Plate Load/Reload: Specifies whether to return the source plate to the loading stack during the print process.

Protocol 10.4: Processing Protein (Serum-Based) Microarrays

How It Works

A previously printed microarray slide is first blocked to reduce nonspecific adsorption. After rinsing, the serum sample is applied to the slide and incubated to capture protein antigens. Next, a labeled secondary antibody is applied that specifically binds to the captured antigen. The slide is finally rinsed, dried, and read.

Required Materials

Printed microarray slide (Custom Made for Application, see publications)
Centrifuge (ArrayIt, Sunnyvale, CA, Cat. No. MHC)
Microarray slide washer (ArrayIt, Sunnyvale, CA, Cat. No. HTW)
Slide scanner (ArrayIt, Sunnyvale, CA, Cat. No. 700A)
Pipette

Reaction cassette (ArrayIt, Inc., Sunnyvale, CA, Cat. No. AHC1x24) or other suitable container for slide processing depending on the format of the microarrays
Serum sample, for example, 3% fetal calf serum or other spent cell culture media
Secondary antibody, fluorescently labeled (Jackson ImmunoResearch, West Grove, PA, Cat. No. 109-095-127)
BlockIt blocking buffer (ArrayIt, Sunnyvale, CA, Cat. No. PA, BKT or BKTHT)
PBS (GIBCO, Rockville, MD, Cat. No. 20012043)
Tween 20 (Sigma, Cat. No. P2287)
PBST (0.05% Tween 20 in 1 × PBS)

Reagent Preparation

1–4000 dilution of Cy3 conjugated goat anti-human IgG/IgM Jackson ImmunoResearch (West Grove, PA) in 1 × BlockIt.

Step-by-Step Protocol

1. Number the printed microarray slide for tracking; some microarray substrates come bar-coded (Note 1).
2. Place microarrays in the slide rack supplied with the wash station.
3. Fill wash station with 400 mL of BlockIt HT, and place microarrays in the wash station.
4. Incubate microarrays in BlockIt buffer for at least 1 h at room temperature (Note 2).
5. Remove slide from blocking buffer, and wash two times in PBST for 15 min each.
6. Spin-dry each microarray for 5 s in a microarray centrifuge.
7. Place microarrays in reaction cassette or similar device.
8. Dilute serum sample in BlockIt buffer, 1:150 v/v.
9. Apply diluted serum sample to microarrays (Note 3).
10. Incubate slide for 1 h at room temperature.
11. Remove slide from reaction cassette, and wash three times in PBST for 5 min each in an HTW (Note 4).
12. Reinsert microarrays in reaction cassette.
13. Apply labeled secondary antibody solution to each microarray (Note 5).
14. Incubate slide for 1 h at room temperature.
15. Remove slide from cassette, and rinse with BlockIt buffer.
16. Wash one time in PBST for 10 min.
17. Wash slide two times in PBST for 5 min each.
18. Wash slide in DDI water for 10 s.
19. Centrifuge microarray 10 s at room temperature to dry slide (Note 6).
20. Scan slide or store in a light proof slide storage box, and keep dry.

Key References

Cai, H. et al. (2005). Development of a novel protein microarray method for serotyping *Salmonella enterica* strains. *Journal of Clinical Microbiology*, 43(7): 3427–3430.

Flinterman, A. E. et al. (2008). Peanut epitopes for IgE and IgG4 in peanut-sensitized children in relation to severity of peanut allergy. *Journal of Allergy and Clinical Immunology*, 121(3): 737–743.

Ho, P. P. et al. (2006). Tolerizing DNA vaccines for autoimmune arthritis. *Autoimmunity*, 39(8): 675–682.

Hueber, W. et al. (2005). Antigen microarray profiling of autoantibodies in rheumatoid arthritis. *Arthritis and Rheumatism*, 52(9): 2645–2655.

Methods and Compositions for Diagnosis and Immunotherapy of Pollen Allergy. U.S. Patent 20070183978.

Ousman, S. S. et al. (2007). Protective and therapeutic role for alphaB-crystallin in autoimmune demyelination. *Nature*, 448(7152): 474–479.

Troubleshooting Guide

Note 1: Track Microarrays

Use a diamond scribe to number slides on the label side. Make a hydrophobic outline around the spots of the microarray with a hydrophobic pen or microarray imprinter.

Note 2: Wash/Block Processing

Once the printing process is complete, and probe samples are coupled to the surface, block a microarray slide surface by incubating for 1–24 h at room temperature in 1× BlockIt buffer. If blocked overnight, it is best to conduct the reaction at 4°C. BlockIt will couple to unreacted binding groups on the microarray surfaces and prevent background fluorescence. Perform this reaction under a coverslip, using 4 μL of BlockIt for every square centimeter (cm^2) of coverslip area. A hybridization cassette should be used to keep the blocking reaction from drying out. Protein binding to the SuperEpoxy slide (ArrayIt, Inc.) surface is extremely stable, and the microarrays can be washed, blocked, and reacted without much loss of coupled protein.

Note 3: Incubation of Serum Test Sample

Put 4 μL of diluted serum (test sample) onto the microarray slide directly on top of the array spots for every square centimeter of reaction area. Common dilutions are 1:150 in BlockIt buffer. If the hydrophobic outline was done properly, then fluid will stay inside the frame. Incubate slides for 1 h at room temperature. With the proper reaction tool, gentle agitation can be provided.

Note 4: First Wash

Use a high-throughput wash station or other suitable washing mechanism that provides agitation.

Note 5: Incubate Labeled Antibody

You may place slides on a tray for incubation with the printed side facing up. Place 4 µL diluted 2° Ab (fluorescent marker) per square centimeter of the slide surface area on the printed side directly over the microarray. The fluorescent markers are photosensitive, so cover with foil during the reaction to prevent photobleaching. Incubate slides for 1 h at room temperature. Gentle agitation can be beneficial.

Note 6: Dry Microarrays

Make sure centrifuge is balanced. Specific tools such as a microarray centrifuge can also be used; they can dry slides and make them ready for scanning in a few seconds. It is preferred that microarrays be scanned immediately to avoid drying of artifacts and photobleaching.

Protocol 10.5: *Ex Situ* DNA Microarray Manufacturing and Processing

Protocol 10.5.1: Experimental Design

The full genomes of most model organisms are readily available, and in many cases, such types of microarrays are made via *in situ* microarray manufacturing techniques. *Ex situ* microarrays provide an affordable alternative to making focused microarrays based on a subset of the genome to verify or test results from high-density experiments. Oligonucleotides can be designed for any gene or variant of a gene. The design directive for gene expression microarrays should be as follows:

1. Select 50- to 70-mers from any organism of choice.
2. Oligos should also include the appropriate controls (i.e., different for different organisms).
3. All oligos should map to the 3′ end of cDNAs.
4. All oligos should map to within 1000 nt of the 3′ end of cDNA.
5. All oligos should correspond to the coding strand (protein strand).
6. Avoid sequence repeats and stretches of poly(A), poly(G), poly(C), and poly(T).
7. Avoid extremes of melting temperatures (T_m's; mixed sequence compositions are best).
8. Run or "blast" search for the oligonucleotide against the data base to avoid cross-reactivity (<70% sequence identity).

Similar to protein microarrays, every microarray must have positive and negative controls. Positive controls to validate the microarray are required to confirm the findings of other groups. Controls often include marker spots, which are always present and provide a means for easy spot finding and downstream data analysis. The goal of microarray experimentation is to investigate a large number of genes simultaneously, not to get every spot recognized. Therefore, the microarray is designed such that experiments can generate data that will allow a predictable profile to emerge, which can be used to extract meaningful information. Prior to starting a microarray experiment, many questions have to be answered. The biological question is the starting place, but there are many other questions and the answers vary greatly depending on the application. There are probably no right answers regarding experimental design because there are so many different approaches that all give the same answer. Not all considerations for experimental design are scientific as cost constraints can often play a role in platform selection. However, the experimental design of clinical diagnostic microarrays must allow statistically and biologically valid conclusions to be extracted from the microarray experiment. The microarray type, expected values, number of technical replicates, number of biological replicates, control types, and specification of controls need to be defined. Defining these parameters constitutes most of the current work being done using microarrays, because the clinical diagnostic implementation of microarrays is still in its infancy. This is outside the scope of work of the typical clinician. In brief, gene expression microarrays measure messenger RNA (mRNA) in cells.

Protocol 10.5.2: Sample Preparation

Once oligos are designed after synthesis and reverse-phase cartridge purification, they are most often shipped dissolved in water. To make sample preparation easy, order oligos such that they arrive normalized to 100 μM concentration in water. Then, make them soluble at a final concentration of 50 μM in a 2× printing buffer that will make the oligo soluble, stabilize the sample, provide efficient transfer onto the surface chemistry, and optimize the attachment to the microarray substrate. Unlike peptides, oligonucleotides are very easy to put in solution, and if epoxide and amino surface chemistry is used, both aqueous and hydroscopic printing buffers can be applied (such as 50% DMSO).

Protocol 10.5.3: Setting up the Source Plates for Printing

See Protocol 10.5.4; the guidelines for source plates are much the same as for oligonucleotides.

Protocol 10.5.4: Executing the Print Run

See Protocol 10.3.5; the guidelines and equipment for oligonucleotides are the same.

Protocol 10.5.5: Processing the DNA Microarray for Gene Expression

Microarray Workflow

All gene expression microarrays have one or more samples hybridizing to the microarray. Processing workflow procedures often include RNA amplification and reverse transcription, labeling, prehybridization, hybridization, and stringency wash steps prior to microarray detection. These conditions change depending on the experimental design and platform.

Amplification

Most gene expression microarrays need 5–20 μg of total RNA per microarray to meet the sample labeling and hybridization requirements. Therefore, gene expression analysis of samples from laser capture microdissection, needle tissue biopsies, or other small clinical samples is impossible without amplifying the very low amounts of total RNA that can be recovered from such samples. Linear amplification of RNA from small samples produces sufficient quantities of RNA for labeling and hybridization. It is critical for the amplification technique to maintain a true representation of the gene expression in the original sample. Similar to RT-PCR and other nucleic acid testing techniques that use RNA, there are multiple steps involved in the isolation and preparation of RNA for gene expression microarray analysis. The quality of the RNA is critical, which implies protecting RNA samples from contamination and degradation, purifying RNA and, finally, characterizing the purified RNA, all of which must be accomplished prior to hybridization. Numerous RNA isolation methods have been published, and a variety of RNA isolation kits are available. The key criterion in choosing a method should be achieving a high yield of intact and pure RNA. The presence of DNA in total RNA samples can cause problems in microarray analysis because many labeling methods label both RNA and DNA with equal efficiency. Labeled DNA can hybridize with microarray targets and can lead to high-level hybridization signals that are not derived from transcripts. RNA cannot be quantitated separately from DNA, so an accurate estimation of the amount of RNA from contaminated RNA preparations is impossible. Therefore, it is advisable to treat total RNA preparations in order to remove DNA contaminants before using the RNA for labeling. This can be achieved with DNase I treatment or by using CsCl gradient centrifugation to separate RNA from DNA.

General recommendations for preparing RNA for microarray analysis are as follows:

1. Minimize the degradation of RNA at all handling stages.
2. Choose an RNA purification method that gives good yields of pure and intact RNA from your samples, even if this means using a complicated protocol.
3. Measure the amount of RNA before using it for microarray labeling.
4. Verify the quality of the RNA before using it for microarray labeling.
5. If possible, purify mRNA for use in microarray analysis.
6. Prepare all the samples for a given set of experiments with the same protocol.

Labeling

Direct labeling incorporates modified nucleotide triphosphates (NTPs) into RNA amplification products during the IVT step of the amplification process. To make amplified RNA (aRNA) that is labeled with fluorescent dyes, a mixture of dye-modified and unmodified (or unlabeled) nucleotides is typically used in order to obtain an optimal ratio of dye-labeled to unlabeled nucleotide for maximum fluorescence. Usually ~200–400 μM of dye-labeled cytidine triphosphate (CTP) is used with 1–3 mM unlabeled NTPs. In general, labeled nucleotides are not incorporated as efficiently as unlabeled molecules during amplification and, therefore, direct labeling does compromise sample yield. If two dyes are used as in a two-color microarray in a direct labeling reaction, it is likely that one dye will not incorporate as well as another, and corrections during data analysis may be required. In direct labeling, the amount of dye incorporated depends on the length of the transcript. Indirect labeling incorporates aminoallyl UTP into amplification products during the IVT reaction. The aminoallyl-modified aRNA produced is covalently coupled to a detectable moiety such as a fluorescent dye or biotin. This method, though more time consuming than direct labeling, can result in very highly labeled aRNA because aminoallyl-modified UTP is incorporated very efficiently by T7 RNA polymerase. The amount of incorporated dye is not transcript length dependent.

Prehybridization

There are three main phases of microarray processing: prehybridization, hybridization, and posthybridization. These steps are optimized to provide the highest possible specificity, signal-to-noise ratio, and reproducibility. Prehybridization prepares the microarray for hybridization. The steps used in prehybridization greatly depend on the sample type, surface chemistry, and buffer system used to manufacture the microarray. However, regardless of the manufacturing specifics, prehybridization processing typically includes blocking, washing, and drying steps. Microarray surface chemistries will bind nucleic acids with high efficiency. Prior to hybridization, the free reactive groups on the surface of the microarray used to bind the target nucleic acid that have not been used must be blocked (deactivated). If not blocked, then nonspecific binding of labeled probe to the surface will take place. This will reduce the probe and compromise the signal-to-noise ratio. Microarrays are blocked and washed with various chemical formulations that typically include salts, detergents, and blocking agents. Some microarray vendors provide microarrays with the prehybridization steps completed. Any excess target nucleic acid must be washed away from the surface prior to hybridization. Excess target nucleic acid in solution will compete with the target bound to the microarray and decrease the measured fluorescence signal from the microarray.

Hybridization

In microarray hybridization reactions, fluorescently labeled fragments of DNA called *the probe* are applied to the immobilized complementary targets on the microarray to form duplexes that will generate a signal read by a scanner. Single-stranded DNA on the microarray is a requirement for hybridization. Ensuring that the target is single stranded is either handled in the manufacturing process or in the prehybridization step. If the target nucleic acid is double stranded, the buffer used to dilute the target in the manufacturing process or during boiling is used to denature the target prior to hybridization. The number of hybridization duplexes generated from the reaction reflects the number of each specific fragment in the probe. The amount of immobilized target nucleic acid must be in excess, and it should not limit the hybridization kinetics. One or more labeled fragment with different fluorescent dyes are often simultaneously hybridized to a single microarray. Specially formulated hybridization buffers are made to stabilize variations in pH; detergents to ensure a clean and homogenous mixture; and other components to increase kinetics, exclude volume, and lower the melting temperature of nucleic acids. Typical microarray hybridization buffers are aqueous based or formamide based. Aqueous hybridizations are done at 65°C, and 50% formamide is used to decrease the melting temperature to 42°C. Hybridization reactions take place at elevated temperatures and over many hours. It is critical that microarrays not be allowed to dry out and a homogenous layer of hybridization mixture be applied to the microarray evenly for the duration of the reaction. The manual tools and automated machines used for microarray hybridization are specifically designed to accomplish these tasks.

Stringency Wash

The purpose of posthybridization washes is to remove all unattached and loosely bound probe molecules from their complementary targets and from the surface of the microarray. In order to prevent false-positive signals, the wash must be stringent. However, it must be optimized so as not to strip the hybridized probe. The stringency of postprocessing will affect the amount of duplexes formed on the microarray during analysis. In general, both automated machines and manual washes are done in SSC–SDS buffer formulations of various concentrations. Typically, three posthybridization washes are performed. The first buffer used, having a high salt content (typically, 1–2× SSC/0.1% SDS buffer), removes most of the hybridization buffer components. The secondary washes are performed with buffers having low salt content (typically, 0.1× SSC/0.1% SDS), and they remove the loosely bound probe from the microarray. Some final washes are done using water. It is critical to check the manufacturer's protocols for exact wash formulations and procedures, as they vary depending on the manufacturing parameters of the microarray. Posthybridization microarrays using fluorescent probes are light sensitive at this stage, so they should be stored in the dark as much as possible prior to scanning. After the final wash step, microarrays should immediately be dried by centrifugation or nitrogen steam. Hybridized microarrays must be stored in the dark and scanned as soon as possible.

Microarray Scan Output

Gene expression microarrays are scanned, and 16-bit gray-scale raw data are generated and saved as a tagged image format file. Scanners are set to saturate only the highest signals on the microarray. This ensures that the most dynamic range of expression can be measured. Many commercial and academic software programs have been designed for quantitation. The intensity of each spot on the microarray is indemnified and verified for quality. Then, spots are typically corrected by subtracting the immediate surrounding background. Next, the corrected intensities are normalized for each spot using various algorithms to ensure the integrity of the data.

Protocol 10.6: The H25K Master Protocol

How It Works

The H25K gene expression microarray (ArrayIt, Sunnyvale, California) is a slide printed with ~25,000 oligonucleotide probes representative of the human genome. This protocol presents a detailed description of preparing a sample for analysis using the H25K microarray as an example. First, total RNA is extracted from tissue. The mRNA is isolated, purified, and converted to aminoallyl cDNA (aaDNA). The aaDNA is labeled with Cy3/5-NHS and hybridized to the DNA microarray.

Protocol 10.6.1: RNA Isolation from Tissue

1. To deep-frozen tissue, add 10 µL of TriReagent per 1 mg of tissue.
2. Sonicate the tissue at 40% maximum power for three 5 s bursts, with 20 s rests between bursts.
3. Leave at room temperature for 5 min, then spin in 4°C centrifuge at 12,000× g for 15 min.
4. Remove the aqueous layer (top fraction) to a fresh, nuclease-free tube.
5. Add 100 µL 1-bromo-3-chloropropane (BCP), and mix vigorously.
6. Leave at room temperature for 5 min, then spin in 4°C centrifuge at 12,000× g for 15 min.
7. Remove the aqueous layer (top fraction) to a fresh, nuclease-free tube.
8. Add isopropanol at half the original TriReagent volume, and mix well.
9. Leave at room temperature for at least 20 min, then spin in 4°C centrifuge at 12,000× g for 15 min.
10. Remove supernatant from tube by pouring it away from the pellet.
11. To the pellet, add 500 µL of 75% EtOH and vortex lightly to wash (do not disrupt the pellet); then spin in 4°C centrifuge at 12,000× g for 15 min.
12. Being careful not to disturb the pellet, remove the supernatant from tube by pouring it away from the pellet, and allow the tube and pellet to dry by inverting the tube onto a clean paper towel for 10–15 min.
13. Resuspend the pellet in 100 µL DEPC H_2O.
14. Transfer the tube to a water bath set at 58–60°C and heat for 5 min.
15. Proceed to RNeasy MinElute cleanup (Protocol 10.6.2).

Protocol 10.6.2: RNeasy MinElute Cleanup

1. Add 350 µL of buffer RLT to the RNA sample in step 15 of Protocol 10.6.1.
2. Add 250 µL of 100% EtOH; mix by pipetting.
3. Apply to MinElute column.
4. Centrifuge for 15 s; discard flow-through.
5. Transfer column to a fresh tube.

6. Add 500 μL buffer RPE.
7. Centrifuge for 15 s; discard flow-through; use the same collection tube.
8. Add 500 μL 80% EtOH (freshly made).
9. Centrifuge for 2 min; discard flow-through and tube.
10. Transfer column to fresh tube.
11. Centrifuge for 5 min with lid open.
12. Transfer column to a fresh 1.5 mL (final collection) tube.
13. Add 14 μL water (at 50°C) to center of minicolumn; wait for 1 min.
14. Centrifuge for 1 min.
15. Read OD on a NanoDrop spectrophotometer; adjust volume to give ~1 μg/μL.

Protocol 10.6.3: cDNA/senseRNA Preparation

Protocol 10.6.3.1: SenseAMP Procedure for First Strand cDNA Synthesis

1. In a microtube on ice, prepare the RNA–primer mix:
   a. 1–7 μL total RNA, 250 ng
   b. 2 μL SenseAMP dT24 RT primer (50 ng/mL) (Vial 1)
   c. 2 μL random primer (250 ng/μL) (Vial 2)
   d. 0–6 μL nuclease-free water (Vial 10) to bring the volume to 11 mL
2. Heat the RNA–primer mix to 80°C for 10 min. Ice immediately for 2 min. Briefly microfuge, and return to ice.
3. For each reaction, prepare a master mix (9 μL total volume) in a separate tube on ice:
   a. 4 μL 5× first strand buffer (or equivalent 5× buffer)
   b. 2 μL 0.1 M DTT (if supplied with enzyme; otherwise, use nuclease-free water)
   c. 1 μL SUPERase In (Vial 4)
   d. 1 μL dNTP mix (Vial 3)
   e. 1 μL Superscript II (or equivalent reverse transcriptase enzyme)
4. Add the master mix to the RNA–primer mix for a volume of 20 μL. Mix gently and microfuge.
5. Incubate at 42°C for 2 h.
6. Microfuge briefly. Add 80 μL of 1× TE buffer to adjust the volume to 100 μL.

Protocol 10.6.3.2: Purification of cDNA

Purify 100 μL of cDNA using the MinElute PCR purification kit (Qiagen, Cat. No. 28006) as follows:

1. Add 500 μL buffer PB to the 100 μL cDNA sample, and mix.
2. Apply the cDNA mixture to the MinElute column and centrifuge for 1 min at 10–14,000× g (~13,000 rpm) in a conventional tabletop microcentrifuge.

3. Discard the flow-through.
4. Place the MinElute column into the same collection tube. Add 750 μL buffer PE to the MinElute column, and centrifuge for 1 min.
5. Discard the flow-through.
6. Place the MinElute column back into the same collection tube.
7. Add 500 μL 80% ethanol to the MinElute column, and centrifuge for 2 min.
8. Disard the flow-through.
9. Place the MinElute column back into the same collection tube. Open the column caps, and place in a microfuge with the cap opposite to the direction of rotation of the rotor to avoid breaking the cap off. Centrifuge for 5 min.
10. Place the MinElute column into a clean, labeled 1.5 mL microfuge tube. To elute cDNA, add 10 μL buffer EB to the center of the column membrane.
11. Incubate at room temperature for 2 min. Centrifuge for 2 min.
12. Discard the column, and save the 10 μL eluted cDNA.
13. If the eluted cDNA is less than 10 μL, bring the volume to 10 μL using nuclease-free water (Vial 10).

Protocol 10.6.3.3: Tailing of First Strand cDNA

1. Heat the purified cDNA (10 mL) to 80°C for 10 min. Ice immediately for 2 min. Briefly microfuge, and return to ice.
2. For each reaction, prepare a master mix (10 μL total volume) in a separate tube on ice:
   a. 2 μL 10× reaction buffer (Vial 6)
   b. 2 μL nuclease-free water (Vial 10)
   c. 4 μL 10 mM dTTP (Vial 5)
   d. 2 μL TdT enzyme (Vial 7)
3. Combine the master mix and the cDNA to a volume of 20 μL. Mix gently, and microfuge.
4. Incubate in a 37°C heat block for 2 min. Do not exceed 2 min.
5. Stop the reaction immediately by heating to 80°C for 10 min. Briefly microfuge, and cool to room temperature (20–25°C) for 1–2 min.

Protocol 10.6.3.4: T7 Promoter Synthesis

1. Add 2 μL of T7 Template Oligo (Vial 8) to the tailed cDNA for a volume of 22 μL. Briefly vortex, and microfuge.
2. Incubate at 37°C for 10 min to anneal the strands.
3. For each reaction, add the following components to a volume of 25 μL:
   a. 1 μL 10× reaction buffer (Vial 6)
   b. 1 μL dNTP mix (Vial 3)
   c. 1 μL Klenow enzyme (Vial 9)

4. Mix gently, and microfuge.
5. Incubate at room temperature (20–25°C) for 30 min.
6. Stop the reaction by heating to 65°C for 10 min. Briefly microfuge, and place on ice for 2 min.
7. Proceed to the *In Vitro* Transcription reaction using half (12.5 μL) of the promoter-modified cDNA. Save the remaining modified cDNA at –20°C for future use or for use in a parallel amplification reaction.

Protocol 10.6.3.5: *In Vitro* Transcription

1. Incubate 12.5 μL of cDNA at 37°C for 10 min to reanneal the strands.
2. Thaw the T7 nucleotide mix (Vial 11) and 10× T7 reaction buffer (Vial 12) at room temperature, and keep at room temperature until used.
3. Thoroughly vortex the 10× T7 reaction buffer (Vial 12) to avoid precipitation of certain buffer components.
4. For each reaction, add the following components at room temperature (20–25°C), in the order listed, to a final volume of 25 μL:
   a. 8.0 μL T7 nucleotide mix (Vial 11)
   b. 2.5 μL 10× T7 reaction buffer (Vial 12)
   c. 2.0 μL T7 enzyme mix (Vial 13)
5. Mix gently, and microfuge. Incubate at 37°C as follows:
   a. 6–16 h in a thermal cycler (with heated lid) set to 37°C.
   or
   b. 5 min in a 37°C heat block, then 6–16 h in a 37°C air hybridization oven
   Note: It is essential to avoid evaporation and condensation of the reaction during this step.
6. Stop the reaction by placing samples at –20°C until ready to proceed with the next step.

Protocol 10.6.3.6: Purification of senseRNA

This protocol is based on use of the RNeasy MinElute Kit (Qiagen, Cat. No. 74204). If necessary, bring each sample to be purified to 100 μL by adding the appropriate volume of RNase-free water supplied with the kit.

1. Add 350 μL of RLT buffer (no BME added), and mix well by pipetting up and down.
2. Add 250 μL of 95–100% ethanol, and mix well by pipetting up and down.
3. Transfer each sample into an RNeasy spin column.
4. Centrifuge at ≥12,000 rpm for 15 s.
5. Discard the flow-through and collection tube.
6. Place each column into a new collection tube supplied with the kit.

7. Add 500 μL of RPE buffer (prepared with ethanol as indicated by Qiagen) to each spin column.
8. Centrifuge at ≥12,000 rpm for 15 s.
9. Discard the flow-through, and place each spin column back in the same collection tube.
10. Add 500 μL of 80% ethanol to each spin column.
11. Centrifuge at ≥12,000 rpm for 2 min.
12. Discard the flow-through and collection tube.
13. Place each spin column in a new collection tube supplied with the kit.
14. Centrifuge spin columns with lids open for 5 min at ≥12,000 rpm to dry and remove any residual ethanol.
15. Place each spin column into a 1.5 mL collection tube supplied with the kit.
16. Add 12 μL of 50°C RNase-free water to each spin column, taking care not to touch the column reservoir with the pipette tip.
17. Incubate at room temperature for 2 min.
18. Centrifuge at ≥12,000 rpm for 1 min.

Protocol 10.6.3.7: Quantitation of senseRNA

Quantitate the senseRNA using a spectrophotometer or other instrument such as a bioanalyzer. Calculate the A260/280 ratio to determine RNA purity. A ratio of 2.0–2.3 is most desirable. Higher ratios may indicate that an excessive poly(A) tail was generated during the amplification reaction.

Protocol 10.6.4: cDNA Synthesis and Indirect Aminoallyl Fluorescent Labeling Kit

How It Works

This kit is optimized for use with 1–2 μg of purified mRNA or 2–10 μg of amplified sense strand mRNA as starting material. The process includes the following steps:

1. cDNA synthesis from mRNA
2. Hydrolysis of RNA
3. Purification of aminoallyl-labeled cDNA
4. Dye binding with aminoallyl-labeled cDNA
5. Purification of dye-labeled cDNA

If used properly, this kit will generate 200–1000 ng of dye-labeled cDNA for hybridization with microarray, depending on the starting material.

Required Materials

Kit contents: 20 labeling reactions for 10 microarrays

Item	Packaging	Storage
Random primer	44 μL in 0.5 mL tube	–20°C freezer
dTVN primer	22 μL in 0.5 mL tube	–20°C freezer
Enzyme mix 4	33 μL in 0.5 mL tube	–20°C freezer
Reaction buffer 4	50 μL in 0.5 mL tube	–20°C freezer
AA-dNTP mix	90 μL in 0.5 mL tube	–20°C freezer
DTT	50 μL in 0.5 mL tube	–20°C freezer
Synthesis stop solution	250 μL in 0.5 mL tube	–20°C freezer
Denaturing solution	250 μL in 0.5 mL tube	–20°C freezer
Neutralization solution	250 μL in 0.5 mL tube	–20°C freezer
Dye-binding buffer	500 μL in 2 mL tube	–20°C freezer
Binding stop solution	150 μL in 0.5 mL tube	–20°C freezer
DNA-binding buffer	6.6 mL in 8 mL bottle	RT (20–25°C)
DNA wash buffer concentrate	1.5 mL in 8 mL bottle	RT (20–25°C)
DNA micro column	30	RT (20–25°C)
Elution tube (1.5 mL)	30	RT (20–25°C)
2 mL wash tube	30	RT (20–25°C)
Nuclease-free water	1.5 mL in 2 mL tube	–20°C freezer

Items Needed, but Not Supplied

Cy3 and Cy5 monoreactive dye packs (Amersham, Cat. No. PA23001/PA25001)
Ethanol 100%
Ethanol 80%
Ice for incubation
PCR tubes

Equipment Needed

Thermal cycler
Microfuge capable of 10,000 rpm or more
Micropipettors
Vortex mixer
SpeedVac

Note: Spin all tubes briefly before opening.

Step-by-Step Protocol

cDNA Synthesis from mRNA

1. In a PCR tube on ice, add the following:
   a. 1–2 μg of mRNA or 2–10 μg of amplified RNA
   b. Random primer (1 μL)
   c. dTVN primer (1 μL)
   d. Nuclease-free water to final volume of 10.5 μL
2. Use both dTVN and random primers when working with purified mRNA.

3. Use 2 μL random primer when working with amplified sense strand mRNA.
4. Incubate at 80°C for 5 min, and allow cooling at room temperature (20–25°C) for 10 min.
5. Briefly spin to collect all the liquid, and add the following:
   a. Enzyme mix 4 (1.5 μL)
   b. Reaction buffer 4 (2 μL)
   c. AA-dNTP mix (4 μL)
   d. DTT (2 μL)
6. Incubate at 42°C for 90 min.
7. Briefly spin to collect all the liquid.
8. Add 10 μL of cDNA synthesis stop solution to the mixture.
9. Store at –20°C, or proceed to step 10.

Hydrolysis of RNA

10. Add 10 μL of denaturing solution to the mixture.
11. Incubate at 65°C for 30 min.
12. Add 10 μL of neutralization solution to the mixture.

Purification of Aminoallyl-Labeled cDNA

13. Before starting, add 6 mL of 100% ethanol to DNA wash buffer concentrate.
14. Add 200 μL of DNA-binding buffer to the mixture, and load onto a DNA micro column.
15. Load the entire contents of the cDNA sample (prepared from step 12) onto the DNA micro column, and place the column in a 2 mL wash tube.
16. Centrifuge for 1 min at full speed in a microfuge (>10000 rpm).
17. Discard flow-through, and reuse the wash tube.
18. Add 100 μL of DNA wash buffer to the column, and spin at full speed for 1 min.
19. Add another 100 μL of DNA wash buffer to the column, and spin at full speed for 2 min. Note: Longer centrifugation time may be necessary to make the column free of ethanol.
20. Discard the 2 mL wash tube, and carefully place the column in an elution tube.
21. Add 10 μL of nuclease-free water that has been prewarmed to 60°C to the center of the column.
22. Wait for 2 min, and spin at full speed for 1 min.
23. Add 10 μL of nuclease-free water that has been prewarmed to 60°C to the center of the column.
24. Wait for 2 min, and spin at full speed for 1 min.
25. Discard the column, and dry the 20 μL cDNA sample in a SpeedVac for 30 min at 42°C.
26. Store the purified cDNA pellet vacuum-sealed at –80°C until ready to perform the dye-coupling step.

Dye Coupling with Aminoallyl-Labeled cDNA

27. Take the dry cDNA samples from the –80°C storage.
28. Add 2.5 μL of nuclease-free water to each cDNA sample.

29. Dissolve the Cy3 or Cy5 monoreactive dye in 45 μL DMSO, and mix by vortexing.
30. To each 2.5 μL cDNA sample, add 4.5 μL of dye-binding buffer.
31. To each 7.0 μL sample, add 3 μL of monoreactive Cy3 or Cy5 dye for a final reaction volume of 10.0 μL.
32. Incubate the 10 μL samples at room temperature for 60 min.
33. Quench each 10 μL sample by adding 6 μL of binding stop solution, and mix well.
34. Incubate at room temperature for 15 min.

Purification of Dye-Labeled cDNA

35. Mix both the Cy3 and Cy5 reactants in a single tube.
36. Add 200 μL of DNA-binding buffer to the mixture, and load onto a DNA micro column.
37. Load the entire contents (prepared in step 35) onto a DNA micro column, and place the column in a 2 mL wash tube.
38. Centrifuge for 1 min at full speed in a microfuge (>10000 rpm).
39. Discard flow-through, and reuse the wash tube.
40. Add 100 μL of DNA wash buffer to the column, and spin at full speed for 1 min.
41. Add another 100 μL of DNA wash buffer to the column, and spin at full speed for 2 min. Note: Longer centrifugation time may be necessary to make the column free of ethanol.
42. Discard the 2 mL wash tube, and carefully place the column in an elution tube.
43. Add 10 μL of nuclease-free water that has been prewarmed to 60°C to the center of the column.
44. Wait for 2 min, and spin at full speed for 1 min.
45. Add 10 μL of nuclease-free water that has been prewarmed to 60°C to the center of the column.
46. Wait for 2 min, and spin at full speed for 1 min.
47. Discard the column.
48. Recover the eluted labeled cDNA for hybridization.
49. Quantify the cDNA, and determine the efficiency of labeling using a spectrophotometer (e.g., NanoDrop).
50. Maintain the labeled cDNA sample at 4°C in the dark till prehybridization.
51. Alternatively, dry the 20 μL cDNA samples in a SpeedVac for 30 min at 42°C.
52. Store the purified cDNA pellet, vacuum-sealed at −80°C.

Protocol 10.6.5: H25K Hybridization of Labeled cDNA

Prepare the labeled cDNA in hybridization solution. Resuspend 1 μg of labeled cDNA in 18.75 μL of molecular-biology-grade water, followed by the addition of 56.25 μL of prewarmed HybIt® 2 hybridization solution. The entire volume is to be hybridized to the microarray (see Note 1 in the Troubleshooting Guide).

1. Rinse the ArrayIt® hybridization cassette chamber and lid with distilled water. Make sure that salt crystals and other contaminants have been removed.
2. Dry the cassette chamber and lid thoroughly with cleanroom wipes (see Note 2).
3. Verify that the flexible rubber gasket is seated evenly in the gasket channel (see Note 3).
4. Preheat both the hybridization cocktail and the hybridization cassette to 42°C.
5. Add 10.0 µL of DDI water to both hydration grooves inside the cassette chamber (see Note 4).
6. Insert H25K microarray with the spotted array side (DNA side) of the microarray facing upward. Make sure that the slide is seated evenly on the base of the cassette.
7. Pipette the hybridization cocktail evenly across the middle of the H25K microarray reaction area using a volume of 75 µL of the hybridization cocktail (see Note 5).
8. Place the coverslip on the microarray. Immediately place the clear plastic cassette lid on top of the cassette chamber and tighten the four sealing screws.
9. Submerge the cassette into a water bath incubator set to 42°C.
10. Hybridize for 3 h at 42°C (see Note 6).
11. Remove the cassette from the water bath, dry it off, and disassemble (see Note 7).
12. Holding the slide by the barcode, immerse the H25K microarray into wash buffer A (1×) to float off the lifter slip (see Note 8).
13. Wash the slide in 1× wash buffer A at 25°C for 5 min.
14. Wash the slide in wash buffer B for 5 min.
15. Transfer the slide into wash buffer C for 30 s to fully clean it.
16. The microarray slide is then dried with a microarray centrifuge and scanned.

Troubleshooting Guide

Note 1: Hybridization Buffer

HybIt 2 Hybridization Solution (1.25×) is recommended. First, resuspend the labeled cDNA in 1 part molecular-biology-grade water, and then add 4 parts of HybIt 2. Prior to use, prewarm the solution for at least 30 s at 42°C and mix by inverting the tube several times.

Note 2: Contamination Issues

Nitrile gloves must be worn at all times during this process. Hand oils, nucleases, and other contaminants can interfere with the hybridization reaction, which proceeds in an ~10 µm layer between the microarray slide and the coverslip.

Note 3: Gasket

The gasket occasionally pulls free of the cassette chamber during use. If this occurs, reinsert the gasket into the gasket channel by applying light pressure. The gasket must be evenly seated in the gasket chamber to prevent leaks between the chamber and lid. Never attempt to use the hybridization cassette without a properly seated gasket.

Note 4: Humidity Control

Water evaporates during the hybridization reaction, producing conditions of 100% humidity in the cassette chamber. A humid environment prevents evaporation of the hybridization solution between the microarray slide and coverslip during the reaction. A hybridization reaction that dries out will fail.

Note 5: Coverslip Directions

This volume will go under the 24 × 60 mm lifter coverslip provided. Prior to pipetting, the hybridization cocktail, H25K microarray, and lifter slip should be preheated to the hybridization temperature (42°C is the recommended temperature for HybIt 2 buffer). Lifter slip should be put immediately onto microarray, and the hybridization cocktail will sheet under the coverslip rapidly (<2 s). The lifter slip can be positioned at an angle, with one side touching the microarray first to avoid air bubbles. If a small air bubble forms, do not worry; it will exit from the reaction after several minutes of hybridization.

Note 6: Hybidization Time

The standard duration of hybridization reactions using HybIt 2 and ~1 μg labeled cDNA is ~3 h at 42°C. To some extent, hybridization times and temperatures can both be adjusted for specific research applications and labeled cDNA amounts. Typically, 2–12 h at 42°C using HybIt 2 is suitable for most applications.

Note 7: Directions for Using the Hybridization Cassette

1. Once the cassette lid is positioned correctly on top of the cassette base, manually tighten each of the four sealing screws by applying downward pressure and turning the screws in a clockwise manner until turning becomes difficult (3–4 half-turns). Check to see that the rubber sealing gasket is seated correctly in the gasket groove. If the gasket is unevenly seated, remove the lid and reposition the sealing gasket before tightening the four sealing screws.
2. After all four sealing screws have been manually tightened, double-check all the four screws by turning each one clockwise again to make sure that the cassette lid is firmly sealed against the rubber sealing gasket in the cassette base. Screws should be tightened by hand. Do not tighten excessively. Use of tools such as pliers or screw drivers is not required; overtightening can crack the hybridization cassette top.
3. Once a microarray is placed inside the hybridization cassette, do not invert the hybridization cassette at any time before, during, or after the hybridization reaction. Always keep the cassette (clear lid) facing upward. Inverting the cassette can cause the microarray slide and coverslip to adhere to the underside of the cassette lid, leading to a loss of hybridization sample and poor results.
4. Incubate the hybridization cassette with the clear cassette lid facing upward in the water bath incubator for the desired period of time.

5. Place the cassette on the laboratory bench, and manually loosen the four sealing screws by turning each of them in a counterclockwise direction. When the four sealing screws have been loosened completely (3–4 half-turns), remove the clear cassette lid. Under certain circumstances, a slight vacuum will prevent the manual removal of the cassette lid. If this occurs, insert a forceps into the slot at the base of the chamber and apply gentle upward pressure.
6. Place the lid aside, and remove the microarray slide from the cassette chamber. Under certain circumstances, the slide will adhere to the base of the cassette. If this happens, insert a forceps into one of the grooves in the cassette base and detach the slide from the base by applying gentle upward pressure.

Note 8: Removal of Coverslip

Gentle agitation in 1× wash buffer A will cause the coverslip to float free from the microarray surface, ~10 s after submerging the microarray into the wash buffer. If the coverslip does not float free of the microarray surface, apply gentle pressure using a fine forceps to remove it. When using forceps to remove a coverslip, avoid contacting the hybridized surface directly, as scratches can reduce the quality of the data collected. After lifter slip removal, insert H25K into a black rack supplied with the wash station.

Appendix A

Microarray Reagent, Materials, and Equipment Sources

Content refers to physical access to probes including their annotation, i.e., gene identity and DNA sequence. Completion of the **Human Genome Project** now permits the use of synthetic Oligo (oligonucleotide) probes in place of RT-PCR derived cDNA probes to study human gene expression. Oligo Probes are less prone to non-specific or cross-hybridization. Long oligo probes, e.g., 50–70mers are used for gene expression while short probes, e.g., 15mer are used for mutation analysis. cDNA probes are still useful for the construction of microarrays for work with other genomes.

Protein content generally refers to access to antibodies that recognize the protein antigen (i.e., the cognate gene-product). Also includes antibody probes to protein variants due to post-translational modification, e.g., phosphorylated proteins.

Nucleic Acids

- **Clones for cDNA**
  - **www.atcc.org**
  - **www.invitrogen.com**
- **siRNA**
  - **www.qiagen.com**
  - **www.ambion.com**
- **Oligos**
  - **www.biosearchtech.com**
  - **www.operon.com**
  - **www.biosource.com**
  - **www.mwg-biotech.com**
  - **www.illumina.com**
  - **www.invitrogen.com**
  - **www.idtdna.com**
- **Antibodies**
  - **www.southernbiotech.com**
  - **www.rndsystems.com**
  - **www.bdbiosciences.com**
  - **www.piercenet.com**
  - **www.vectorlab.com**
  - **www.abcam.com**
  - **www.invitrogen.com (Zymed)**

Figure A1
Content providers.

SSC (saline-sodium citrate)
20X = 3.0 M NaCI, 0.3M citrate, pH 7.0

SSPE (saline- sodium phosphate EDTA)
20X = 3.0 M NaCl, 0.2M Na H_2PO_4, 0.02M EDTA, disodium, pH 7.4

TAE (*Tris*-acetate EDTA)
10X = 0.4M Tris, 0.01M EDTA, disodium, pH 7.8 adjusted with acetic acid

TBE (*Tris*-borate EDTA)
10X = 0.89M Tris, 0.89M boric acid, 0.02M EDTA, disodium, pH 8.3 adjusted with NaOH

Denaturing Solution
1.5M NaCI, 0.5M NaOH, pH 13.0

Neutralizing Solution
1.5M NaCI, 1.0M Tris, pH 7.5 adjusted with HCI

Denhardt's Solution
50X = 1% Polyvinylpyrrolidone, 1% BSA, 1% Ficoll

Figure A2
Buffers for nucleic acid microarrays.

Oligos	**5 μM to 40 μM**
cDNA	**1 nM to 1 μM**
Proteins	**0.1 to 1 mg/mL**

- Nucleic Acids
  - Variables to consider
    - Synthesis & purification of nucleic acid probes
      - Check with vendor about co-purification issues
      - HPLC salt carryover
    - Sequence & concentration
      - Concantenation
      - Aggregation
- Proteins
    - Non-specific adsorption
      - Quill
      - Substrate
    - Stability issues—denaturation & aggregation
      - pH
      - Salts
      - Temperature
      - Dehydration

Figure A3
Probe 'ballpark' print concentrations.

What features are most important in selecting substrates?

Optical clarity

Flatness

Intrinsic fluorescence

Surface chemistry

Non-specific background

Cost

- **Glass Slides (nucleic acids)**
  - **www.arrayit.com (TeleChem)**
  - **www.eriemicroarray.com (Erie)**
  - **www.cel-1.com (Cel Associates)**
  - **www.corning.com (Corning)**
  - **www.amershambiosciences.com (GE Healthcare)**
- **Glass Slides (proteins)**
  - **www.fastquant.com (Whatman S&S)**
- **Plates**
  - **www.gracebio.com (Grace BioLab)**
  - **www.gbo.com (Greiner Bio-one)**
  - **www.us.schott.com/nexterion (Schott)**

Figure A4
Substrate vendors.

Array Scanners

Slides

GenPix
www.axon.com
ScanArray (GSI Lumonics)
www.perkinelmer.com
GeneMachine UC4
www.genomicsolutions.com
arrayWoRx
www.appliedprecision.com
Agilent (SureScan)
www.agilent.com

Plates

NovaRay
www.alphainnotech.com
LS Reloaded
www.tecan.com

Image Analysis Softwares

ImaGene
www.biodiscovery.com
ArrayVision
www.imagingresearch.com
GenePix Pro
www.axon.com
IPLab (Scanalytics)
www.bdbiosciences.com
Array-Pro (Media Cybernetics)
www.mediacy.com
QuantArray (GSI Lumonics)
www.perkinelmer.com
ScanAlyze(freeware)
Stanford University (Eisen Lab)
rana.stanford.edu/software
Various freeware downloads
The Institute for Genomic Research
www.tigr.org/softlab

Figure A5
Array scanners and image analysis softwares.

Dye Instability: signal degradation for Cy5 has been observed and traced to atmospheric ozone levels. A few ppb ozone can bleach Cy5 while leaving Cy3 unharmed.
Dye Bias: enzymatic incorporation of various dyes into nucleic acids has been found to be not equivalent. Dye swapping experiments are suggested in order to obtain accurate results from dye ratio.
Calibration: slide scanners require calibration in order to assure correct representation of signal intensity for each dye. Laser power decreases with time as do dyes.

- Notes
  - Cy5 or Alexa 647 show intensity loss within 10–30 s in 5–10 ppb ozone
  - Cy3 or Alexa 555 no change at > 100 ppb
  - Keep ozone level in lab < 2 ppb
  - Use HVAC with scrubbers

Figure A6
Precautions in reading microarrays.

Mwt. Oligo (g/mole) = {[**#A's** * 312.2] + [**#C's***288.2] + [**#G's***328.2] + [**#T's***303.2]} **– 61.0**
Mwt. Oligo (g/mole) approximately 330* # Bases
nmoles Oligo = (A260nm * 90) / (# Bases)

(1) OD A260nm	~ 33 µg/mL ss DNA
	~ 50 µg/mL ds DNA
	~ 40 µg/mL RNA
	~ 4.5 nmole, 20mer oligo
	~ 2 nmole, 50mer oligo

1 µg of 1000 bp DNA = 1.52 pmole (3.03 pmoles of ends)
1 pmole of 1000 bp DNA = 0.66 µg
$\mathbf{Tm}_{short}$ (°C) = 2° *(A + T) + 4° *(G + C) for short oligos, ≤ 20mer
$\mathbf{Tm}_{long}$ (°C) = 81.5 + 16.6 log M + 41 (xG + xC) –500/L – 0.62F
where M = Molar salt conc., x = mole fraction, L = oligo length, F = Molar conc. formamide

$$\textbf{Tm (Nearest Neighbor)} = \left\{ \frac{\Delta H^\circ - 1000}{A + \Delta S^\circ + R\ln(C_t / 4)} \right\} - 273.15 + 16.6\log[Na^+]$$

Figure A7
Useful calculations and conversions.

Cells
↓
Extraction
↓
total RNA

1 mg Tissue – Major Organs
High (2–4 μg): Liver, Spleen, Heart
Moderate (0.05–2 μg): Brain, Kidney, Lung
Low (< 0.05 μg): Bone, Adipose

- Kits
  - www.ambion.com
  - www.invitrogen.com
  - www.promega.com
  - www.qiagen.com
- Protocols
  - Boom
- Alternatives
  - Oligo-dT
  - Magnetic beads
  - Phenol-GuSCN/Chloroform (Trizol)

Figure A8
RNA sample extraction kit vendors.

1. Dilute 5′ (or 3′) amino-linked oligonucleotide in print buffer containing MSS to 20 μM final conc.
2. Print probe onto aldehyde slides.
3. Dry for 12 h, 25°C, < 30% rel. humidity.
4. Rinse slide with 0.1% SDS.
5. Rinse slide with ddH_2O.
6. Transfer slide into boiling ddH_2O, 3 min.
7. Remove, plunge into ice cold, 100% ethanol, 30 s.
8. Remove, centrifuge to uniformly dry.

- Slides
  - Protocol developed for TeleChem's SuperAldehyde Slide
    - Recommend use with Micro Spotting Solutions
    - Dilute print buffer with MSS+
  - Slide chemistries vary so check with vendor on recommended protocols
- Buffers
  - A simple print buffer such as sodium phosphate at pH 8 to 9 works well
    - For example, 150 mM sodium phosphate, pH 8.5
  - Additives
    - 0.005% Tween-20
    - 0.001% Savrcosyl
    - 50% DMSO

Figure A9
Oligonucleotide printing protocol.

1. Transfer antibodies from glycerol buffer into glycerol-free PBS solution.
2. Prepare proteins at 0.1–0.3 mg/mL.
3. Transfer 4 μL into 384-well source plate.
4. Array solutions from source plate onto the poly-L-lysine coated glass slide.
5. Rinse microarrays briefly in 3% non-fat milk in PBS, pH 7.4 containing 0.1% Tween 20.
6. Soak slides overnight at 4°C in 3% non-fat milk in PBS, pH 7.4 containing 0.02% sodium azide as preservative.
7. Just prior to use rinse slides at room temperature with PBS.
8. Maintain slides in PBS buffer up until incubation with sample.

- Slides
  - Protocol developed for a poly-L-lysine slide
    - Glycerol causes spreading and may reduce binding to substrate
    - Use spin-columns or microdialysis for smaller sample sizes for buffer exchanges
- Buffers
  - Non-fat milk or casein blocks nonspecific sites on slide
  - Tween 20 serves (2) purposes;
    - Removal of non-specific protein
    - Hydration of bound protein reducing denaturation
- Maintain hydration—do not allow slides to dry out

Figure A10
Antibody printing protocol.

Cleaning and pin inspection are obligatory!

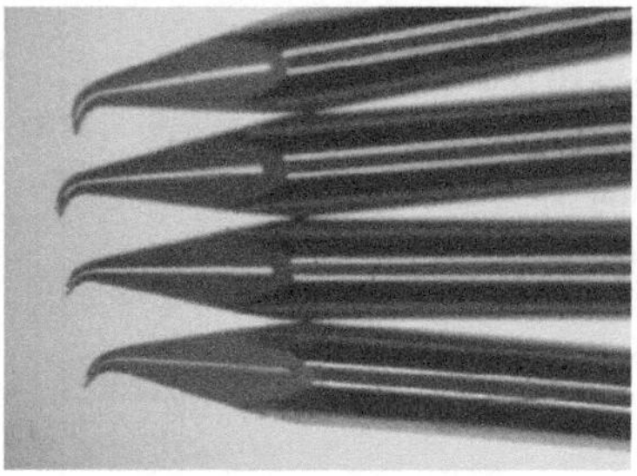

Pre-run print programs!

Figure A11
Quill pin vendors.

Prepare Print Oligos (Probes)
Use HPLC purified, desalted oligos.
Dissolve at stock conc. ~200 µM in ddH_2O, then aliquot.
Dilute to ~5 to 20 µM in a suitable buffer to match slide surface chemistry.
Prepare Source Plate
If possible use a robotic liquid handler to prepare source plate.
Need to map to pin configuration.
Dispense 10–20 µL per well.
Avoid air-bubble formation and centrifuge plate to remove.
Store plate refrigerated (short term) or freeze (longer term) but tightly sealed.

Figure A12
Preparation of print oligo probes.

- **Preparation of Oligonucleotide Microarrays**
  - T_d adjustments
  - Calculation of probe conc.
  - Preparations for printing
  - Printing
  - Post-print workups
  - Quality assessments

Figure A13
Printers.

- Printers
  - Bench Scale
    - www.arrayit.com
      - NanoPrint
    - www.genomicsolutions.com
      - Cartesian
      - GeneMachines OmniGrid
      - BioRobotics MicroGrid
    - www.bio-rad.com
      - VersArray ChipWriter Pro
  - High Throughput
    - www.genetix.co.uk
      - Qarray2
    - www.genexpbiosciences.com
      - BioGrid Arrayer

Appendix B

Image Analysis

The following figures will guide you through the process of obtaining data from a raw image file of a scanned microarray slide. The example given is from Todd Martinsky's H25K gene expression microarray (ArrayIt, Inc.) slide experiment. Images are courtesy of BioDiscovery, Inc., from analysis using the ImaGene™ software, ver. 8.0.

A developed microarray slide is first scanned and the resulting image digitized and saved in an image format (e.g., JPEG, TIFF, GIF) on a disk. Most images are saved as 16-bit gray-scale images in TIFF format. In a 16-bit gray-scale image, the decimal number 65,536 is equivalent to the 16-digit binary number 1111111111111111 and is the highest signal possible on a microarray, whereas zero is the lowest, and this format is used to generate quantitative data. The image analysis software is launched and the image file uploaded or imported. In this case, the slide is an H25K microarray, which has been hybridized with a fluorescently labeled cDNA, that has been scanned and saved with a .tif file extension.

The next step is to import a template file. This permits the software to recognize the dimensions of the image that it will process and select the format to place the data in.

Once the image is formatted, the ~25,000 spots on the slide need to be assigned a spatial (x,y) address. A grid file that specifies the dimensionality of the microarray, the spot diameter, and address is loaded. This information is stored in a file called a GAL file, short for gene array list, and is recognized by the image analysis software. The GAL file also contains all the gene names and other information about the spots on the microarray. The next step is to position the grid over the array image or overlay. Because the spots are aligned on the microarray within 25 μm and may vary slightly in diameter, it is necessary to fine-adjust the spots within the grid using software tools. Most good software programs such as ImaGene do this automatically with little user intervention.

Having accomplished grid placement and spot adjustments, the final step is to compute the pixel intensity values for the spots within the microarray. This is automatically done and the data placed in tabular form so that it may be exported for further analysis. For example, the table can be exported as an Excel file.

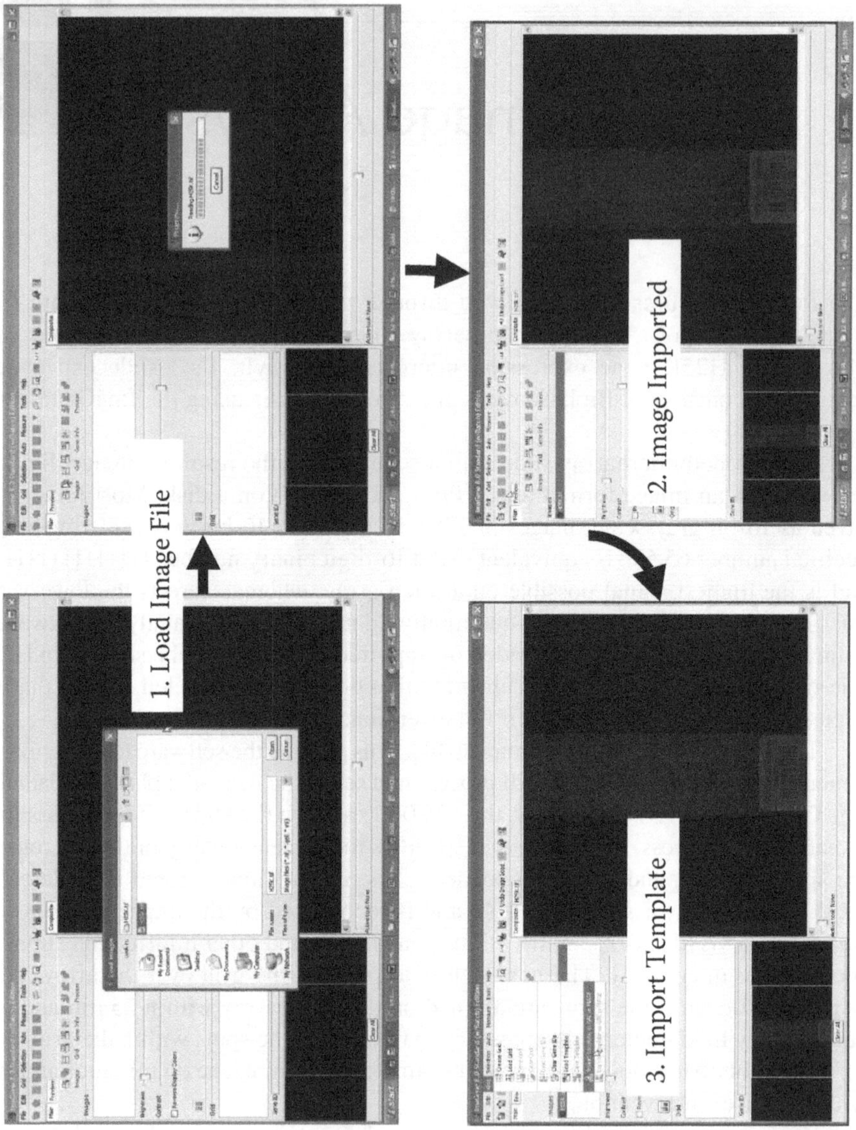
1. Load Image File
2. Image Imported
3. Import Template

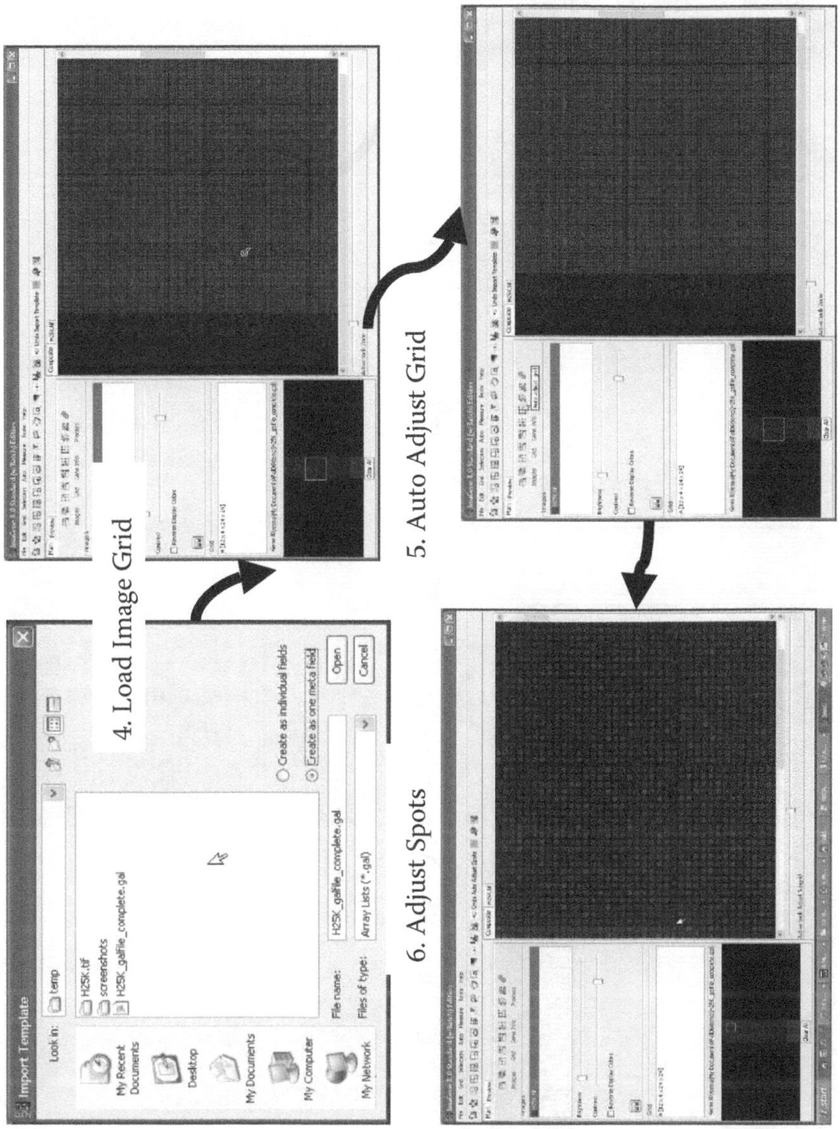
4. Load Image Grid
5. Auto Adjust Grid
6. Adjust Spots
Import Template
Look in:
temp
H25K.tif
screenshots
H25K_galfile_complete.gal
My Recent Documents
Desktop
My Documents
My Computer
My Network
File name:
H25K_galfile_complete.gal
Files of type:
Array Lists (*.gal)
Create as individual fields
Create as one meta field
Open
Cancel

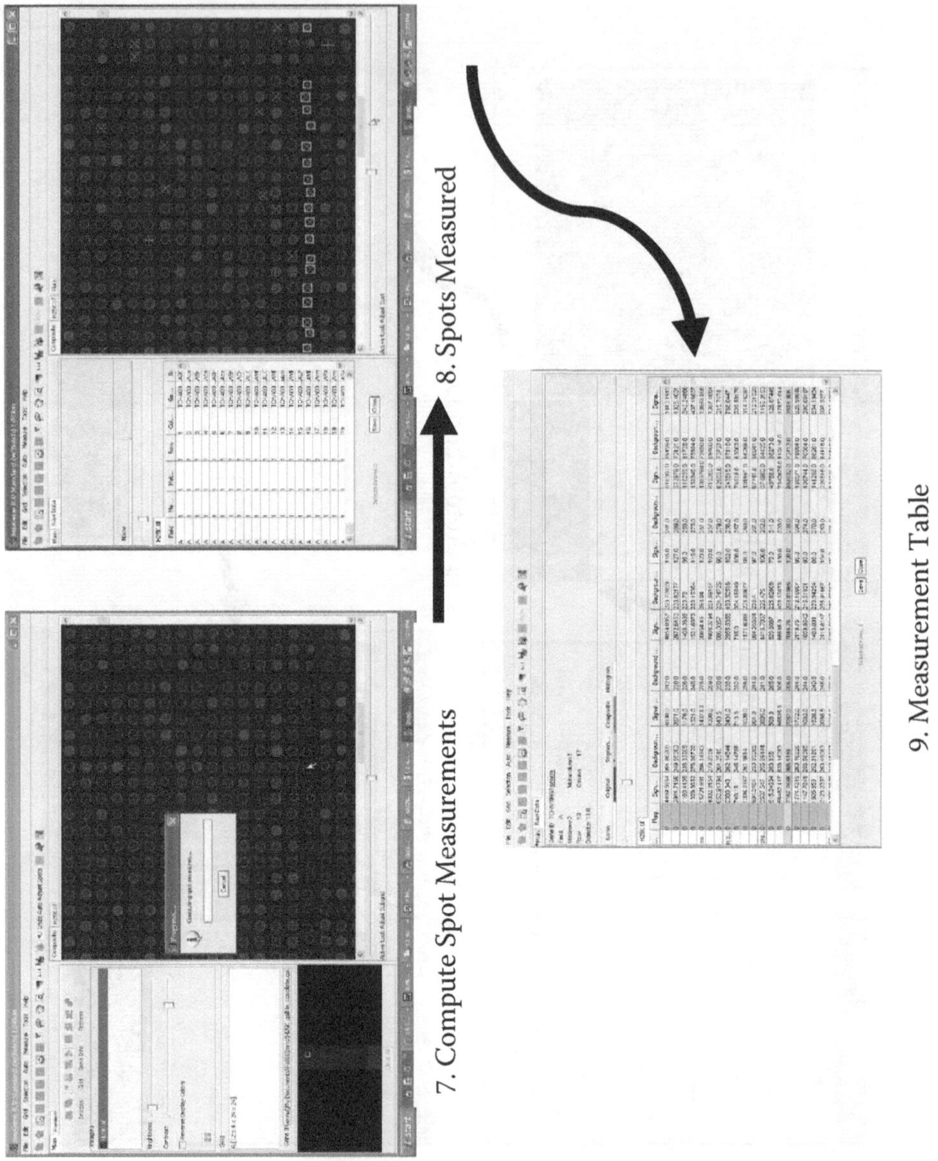
7. Compute Spot Measurements
8. Spots Measured
9. Measurement Table

Index

U

V